Digital Twins for Wireless Networks

Muhammad Khalil Afzal • Muhammad Naeem •
Waleed Ejaz
Editors

Digital Twins for Wireless Networks

Overview, Architecture, and Challenges

Springer

Editors
Muhammad Khalil Afzal
Department of Computer Science
COMSATS University Islamabad
Wah Campus, Islamabad, Pakistan

Waleed Ejaz
Department of Electrical and
Computer Engineering
Lakehead University
Thunder Bay Barrie/Orillia, ON, Canada

Muhammad Naeem
Department of Electrical and
Computer Engineering
COMSATS University Islamabad
Wah Campus, Punjab, Pakistan

ISBN 978-3-031-73678-0 ISBN 978-3-031-73679-7 (eBook)
https://doi.org/10.1007/978-3-031-73679-7

This Springer imprint is published by the registered company Springer Nature Switzerland AG
The registered company address is: Gewerbestrasse 11, 6330 Cham, Switzerland

Acknowledgments

We would like to express our gratitude to everyone who participated in this book and made this book a reality. In particular, we would like to acknowledge the hard work of the authors.

We would also like to acknowledge the efforts of the reviewers, who enabled us to select these chapters out of many we received and help to improve the quality of the chapters by providing their valuable comments.

Lastly, we are very grateful to the editorial team at Springer Nature for their support throughout the publishing of this book. Special Thanks to our editor Susan Evans Grove for her great support and encourgement. We would also like to thank Reshma Begam, Kritheka Elango and Joseph Quatela from Springer Nature for managing and production process of this book.

Introduction

The sixth-generation (6G) communication systems are anticipated to provide network connectivity for an extensive range of use cases in a variety of emerging vertical industries. Consequently, a new set of challenging requirements and more stringent key performance indicators have to be considered, a novel architecture has to be designed, and unique enabling technologies shall be developed to fulfill the technical, regulatory, and business demands of the communication service customers.

6G networks are expected to offer even faster speeds, lower latency, and greater capacity compared to 5G networks, which will enable new applications and use cases that are currently not possible. Improved quality of life by enabling various applications (emerging Internet of everything applications) such as healthcare, brain-computer interactions, and extended reality is the main focus of future wireless services. Quality of experience, latency, and reliability are the key requirements of these applications. To meet these diverse requirements there is a need to assist wireless systems with unique technologies. Self-sustaining wireless systems (intelligence, seamless, and ubiquitous connectivity) and proactive-online-learning-enables systems (Intelligent analytics) are two trends in future wireless systems. The digital twin technology is one of the most promising technologies that can be instrumental in realizing the technical and business objectives of 6G communication systems.

A digital twin is a virtual imitation of a physical object or system. In a wireless system, a digital twin can be used to model and analyze the behavior of the network and its components, such as antennas, transmitters, receivers, sensors, and other devices in wireless networks. One of the key benefits of using a digital twin for a wireless system is that it can help network operators and engineers to optimize the performance of the wireless network by simulating different scenarios and configurations. Other benefits include improve efficiency, cost saving, and enhanced security. In 6G networks, a digital twin could be used to simulate and optimize the performance. This could include simulating different network topologies, testing the performance of different network protocols and algorithms, and optimizing the placement of network infrastructure.

To create a digital twin of a wireless network, a detailed model of the network and its components must be developed, based on real-world data and conditions. This model can then be used to simulate the behavior of the network under different conditions and settings and to visualize the results in real time.

Chapter 1 explores the integration of digital twin networks into the landscape of wireless communication systems, particularly in the context of sixth-generation (6G) wireless networks. The rapid advancements in artificial intelligence, machine learning, and high-performance computing have paved the way for leveraging digital twins as powerful tools for designing, analyzing, and operating wireless networks. Digital twins, as virtual replicas of physical entities, offer real-time monitoring, simulation, analysis, optimization, and prediction capabilities for various applications in smart cities, manufacturing, retail, and beyond. By connecting real-world operations with real-time data streams and leveraging advanced analytics and predictive modeling, digital twin networks enable proactive problem-solving, resource optimization, and decision-making. This chapter discusses the basics of digital twin networks, their working principles, and their relevance to wireless networks. It also highlights the potential challenges and benefits of integrating digital twins into wireless communication systems and outlines future research directions. Overall, digital twin networks present a promising approach for creating intelligent and responsive wireless networks capable of meeting the diverse requirements of emerging applications in the era of 6G.

Chapter 2 covers a broad range of topics within Wireless Body Area Network (WBAN) digital twins, encompassing modeling and simulation of tissue/organ dielectric properties, specific absorption rates, signal path losses, and heat generation due to signal propagation in the human body. This kind of digital twin (DT) was very important in the development of the Internet of Things (IoT) in the field of smart health, such as wireless wearable sensors and wireless implant devices. By DT, the appropriate wearable sensor position and its performance can quickly be identified. On the other hand, the power level of the propagated wireless signal can be initially estimated to optimize the efficiency of the implant devices and the safety of the body tissues/organs. In addition, this chapter proposes a graphical user interface-based simulation framework, which includes dielectric human tissue/organ models and path loss models, for realizing a digital twin ecosystem.

Chapter 3 discusses the emergence of Digital Twin (DT) technology that holds immense promise for revolutionizing modern communication systems, particularly in the context of future Metaverse-based networks. DT involves digitally defining and modeling physical entities, and facilitating feedback optimization through simulation, control, and prediction. While DT has seen success in manufacturing and complex system operation, its application in next-generation (nextG) networks is nascent, facing challenges such as building high-fidelity twins, adapting to dynamic network nature, and integrating with advanced machine learning (ML) models. Digital Twin Networks (DTNs) offer a compelling solution, providing benefits like accelerated network development, enhanced resource utilization, and streamlined management. In a DTN, virtual replicas of physical entities form a comprehensive network system, enabling real-time simulation, monitoring, and analysis. DTNs find applications across network deployment, operation, and design phases, facilitating remote commissioning, diagnostics, and predictive analytics to optimize efficiency and performance. Despite the challenges, the fusion of ML techniques with DTNs offers tangible benefits across resource allocation, channel estimation, and user behavior prediction. This chapter aims to explore the synergy between ML and DT in addressing these challenges and leveraging opportunities, highlighting key insights for DTN-enabled network applications.

Chapter 4 elaborates the sixth-generation applications with intensive computation and low latency requirements that generate the offloading need for mobile users. For optimal offloading, the digital twin is a potential technology in multi-stage networks with mobile users; unmanned aerial vehicles act as mobile edge computing servers and cloud server. In a multi-stage network, the high computation tasks with stringent latency requirements on mobile devices are handled efficiently. This chapter proposes a digital twin-assisted multi-stage network for low-latency communication and computation. An optimization problem for mobile users's task latency minimization is formulated by optimizing the offloading portions in multistages and association with UAV-MECs. The optimization problem is solved through an alternate optimization algorithm, including the learning algorithm and interior point method. The simulation was carried out under different scenarios to emphasize the usefulness of the proposed network.

Chapter 5 discusses the rapid development of wireless technologies, such as the Internet of Things and widely deployed fifth-generation (5G) networks, which are proved to be the embarkment of envisioning, and planning for sixth-generation (6G) mobile networks. It is believed that 6G will provide extremely high data rates, low latency, and improved edge intelligence for hundreds of billions of end devices, connected to the 6G network. Thus, a huge amount of data will be generated from these devices, which requires tremendous computation and communication resources to be provided by edge servers. The gap between users' requirements and edge servers' capability of service provisions in 6G systems is mitigated with digital twin with close monitoring, real-time interaction, and reliable communication between the digital space and the physical systems, which can in turn optimize the running of the physical systems. This chapter entails the basic knowledge of Digital Twin (DT) for 6G Wireless networks. Moreover, the properties of Federated Learning that can enhance the DT for 6G to provide high level performance in communication by presenting the case studies of smart manufacturing and smart cities.

In Chap. 6 challenging applications such as holographic communications, Internet of Everything (IoE), tactile Internet, and quantum communications can be well managed and optimized with the help of digital twins. Digital twins-enabled wireless networks provide the capability of self-configured and online learning to underlying processes and systems, which not only enhances the data rate and connectivity but also maximizes the system security. In this chapter, different security attacks on the digital twins-enabled wireless technology have been explained with a focus on the physical interaction layer and twin layer. Different security measures that could be adopted to mitigate these threats have also been discussed in detail with respect to different layers of communications for the digital twins-enabled wireless networks. Open issues, challenges, and future research directions regarding the security paradigm for digital twins-enabled wireless networks have also been encompassed briefly.

Chapter 7 explains the era of technology and development that is facing major challenges in terms of disasters (floods, earthquakes, tornadoes, and many more) due to global warming. A solution that can sufficiently and technically handle the catastrophic situation where human lives are in danger can be communicated safely with the providence of medical and relief aid. Here, metaverse, an emerging technology, paves its way. Implementation of metaverse has already eased many industries, medical sciences, and office staff, and now its scenario-based deployment can also be helpful in disaster situations. In this chapter, we will provide the collaboration of metaverse technology with Unmanned Aerial Vehicles (UAVs) that will help the affected people get out of danger through visual monitoring. Multiple terminologies, technologies, elements, and tools are associated with the word metaverse. We will briefly review all the concepts mainly focusing on communication technologies and the Digital twin concept. We have integrated the concept of wireless infrastructure less medium, i.e., UAVs, where communication is provided with the help of UAVs.

Chapter 8 explains the applications of the Unmanned Aerial Vehicles (UAVs) in challenging fields such as military, defense, civil, and agriculture that have made them a promising futuristic technology. The futuristic UAVs using the 5G and beyond 5G technology with more sophisticated and intelligent algorithms require a framework for their optimization and troubleshooting. This has invited the researchers to come out with the advanced concept of digital twin for the UAVs. Digital twin enabled UAVs provide a framework for the optimization, training, and testing of the intelligent algorithms using the real-time data from the drone operations. In this chapter, an overview and investigation into the research work related to the digital twin enabled UAVs has been provided with some higher level insights. Digital twin enabled UAVs architecture has been discussed with its physical and virtual layers. Different applications of the UAVs with their major remote sensing monitoring role have been investigated in more detail. Various case studies of the UAVs employing the digital twins for online learning and optimization are also discussed. Open issues, challenges, and future research directions related to the working of the digital twin enabled UAVs have also been highlighted.

Contents

About the Editors

Muhammad Khalil Afzal (SM'16) received his MCS and MS degrees in Computer Science from COMSATS Institute of Information Technology, Wah Campus, Pakistan, in 2004 and 2007, respectively, and his PhD degree from the Department of Information and Communication Engineering, Yeungnam University, South Korea, in December 2014. He served as a lecturer from January 2008 to November 2009 at Bahauddin Zakariya University, Multan, Pakistan, and from December 2009 to June 2011 at King Khalid University Abha, Saudi Arabia. Currently, he is working as an Associate Professor (Tenured) in the Department of Computer Science at COMSATS, Wah Cantt, Pakistan. He has served as a Guest Editor of *IEEE Internet of Things Journal*, Elsevier *Computer Communication*, *IEEE Communication Magazine*, *Transactions for Emerging Telecommunications Technologies* (ETT), *Future Generation Computer Systems* (Elsevier), *IEEE ACCESS*, *Journal of Ambient Intelligence and Humanized Computing* (Springer), *MDPI Sensors Journal* and reviewer for *IEEE ACCESS*, *Computers and Electrical Engineering* (Elsevier), *Journal of Network and Computer Applications* (Elsevier), Future Generation Computer Systems (FGCS), and *IEEE transaction on Vehicular Technology*. He is the recipient of the best paper awards of ACM 7th International Conference on Computing Communication and Networking Technologies and in International Conference on Green and Human Information Technology (ICGHIT-2018)

and a fully funded scholarship for a master's and PhD. He is the recipient of research projects funded by the Higher Education Commission (HEC), Pakistan; the National Grassroots ICT Research Initiative, Ignite, Pakistan; the European Union's Programme Horizon 2020; and the British Council, UK. His research interests include wireless sensor networks, ad hoc networks, Data-driven intelligence in wireless networks, Smart Cities, 5G, and IoT.

Muhammad Naeem received the BS and MS degrees in electrical engineering from the University of Engineering and Technology, Taxila, Pakistan, in 2000 and 2005, respectively, and the PhD degree from Simon Fraser University, Burnaby, BC, Canada, in 2011. From 2000 to 2005, he was a Senior Design Engineer with Comcept (Pvt) Ltd., Islamabad, Pakistan, where he participated in the design and development of smart card-based GSM and CDMA pay phones with the Department of Design. From 2012 to 2013, he was a Postdoctoral Research Associate with the Wireless Networks and Communications Research (WINCORE) Laboratory, Ryerson University, Toronto, ON, Canada. Currently, he is a Professor with the Department of Electrical Engineering, COMSATS University Islamabad, Wah Campus, Wah, Pakistan. He is also a Microsoft-Certified Solution Developer. His research interests include the optimization of wireless communication systems, non-convex optimization, resource allocation in cognitive radio networks, and approximation algorithms for mixed integer programming in communication systems. He was a recipient of the NSERC CGS Scholarship.

Waleed Ejaz (S'12-M'14-SM'16) is an Associate Professor in the Department of Electrical Engineering at Lakehead University, Barrie Campus, ON, Canada. He was an Assistant Professor in the Department of Engineering & Applied Science at Thompson Rivers University, Kamloops, BC, Canada, from Sep. 2018 to Aug. 2020. Previously, he held academic and research positions at Ryerson University, Carleton University, and Queen's University in Canada. He received the BSc and MSc degrees in Computer Engineering from the University of Engineering and Technology, Taxila, Pakistan, and the National University of Sciences and Technology, Islamabad, Pakistan, and a PhD degree in Information and Communication Engineering from Sejong University, Republic of Korea, in 2014. He has co-authored over 90 papers in prestigious journals and conferences and 3 books. His current research interests include the Internet of Things (IoT), energy harvesting, 5G and beyond networks, and mobile edge computing. He is an Associate Editor of the *IEEE Communications Magazine*, *IEEE Canadian Journal of Electrical and Computer Engineering*, and the *IEEE ACCESS*. Dr. Ejaz completed certificate courses on "Teaching and Learning in Higher Education" from the Chang School at Ryerson University. He is a registered Professional Engineer (P.Eng.) in the province of Ontario, Canada.

Chapter 1
Overview of Digital Twin, Architecture, and Applications

Sana Sharif, Muhammad Khalil Afzal, Muhammad Naeem, and Waleed Ejaz

1.1 Introduction

The rapid advancements in artificial intelligence (AI), machine learning (ML), and high-performance computing, driven by graphics processing units (GPUs), have ushered in a new era for digital twins [1]. These digital replicas of physical entities have applications in diverse fields, such as smart cities, manufacturing, and retail. As the research community shapes the landscape of the sixth generation (6G) wireless networks, it is clear that novel applications will emerge, including extended reality, immersive multimedia, holographic communication, autonomous driving, network, and computing convergence, multidimensional sensing, pervasive intelligence, and connectivity for Industry 4.0 and beyond [2]. Addressing the diverse requirements of these 6G use cases demands cutting-edge technologies. Digital twins are an up-and-coming solution for designing, analyzing, and operating 6G wireless networks. The integration of AI further enhances the potential of digital twins, transforming

S. Sharif
Lakehead University, Barrie Campus, Barrie, ON, Canada
e-mail: ssharif@lakeheadu.ca

M. K. Afzal
Department of Computer Science, COMSATS University Islamabad, Wah Campus, Islamabad, Pakistan
e-mail: khalilafzal@ciitwah.edu.pk

M. Naeem
Department of Electrical and Computer Engineering, COMSATS University Islamabad, Wah Campus, Punjab, Pakistan

W. Ejaz (✉)
Department of Electrical and Computer Engineering, Lakehead University, Thunder Bay Barrie/Orillia, ON, Canada
e-mail: wejaz@lakeheadu.ca

M. K. Afzal et al. (eds.), *Digital Twins for Wireless Networks*,
https://doi.org/10.1007/978-3-031-73679-7_1

them into perceptive and cognitive intelligence generators by analyzing massive historical and real-time network data [3]. A twin refers to an entity with a genetic similarity due to its origin from the same source. Digital twins are highly accurate virtual models designed to mirror and represent a physical object in the digital world. But what exactly are they? Consider it a high-tech mirror reflecting the real world into the digital reality. We're talking about a virtual representation of a physical entity; it can be as small as a component, a machine, a process, or even an entire town. From essential elements to complex systems, digital twins in 6G facilitate monitoring, simulation, analysis, optimization, design, and prediction of physical systems' behaviors. However, the promises of digital twins in 6G come with unprecedented challenges. Real-time sensing, coupled with AI technologies, is crucial for driving physical networks continually to an optimal state. The extensive computational capabilities required by digital twins necessitate scaling out physical device computing capacity, leading to split computing as a trend for 6G. The trustworthiness of physical networks supporting digital twins requires adaptive security measures, including physical layer security solutions and distributed anomaly detection. Moreover, the evolution of digital twins to represent the 6G environment and interactions between digital twins demand massive information exchange through terahertz (THz) communication, which plays a crucial role in 6G.

A digital twin network is described as a digital replica of the entire life cycle of a physical network. It utilizes data and models to create a physically accurate network simulation platform, providing up-to-date network status and predicting future network states [4]. Unlike conventional network simulators, digital twin networks support two-way communication between the physical and virtual twin networks, enabling real-time interactive mapping and closed-loop decisions [3]. Digital twins aren't one-size-fits-all. There are product twins, helping design and test cool gadgets. Process twins keep industries ticking like clockwork, while system twins handle the big players like intelligent cities [5].

Hence, integrating digital twin technology into 6G wireless networks brings immense potential for transformative applications. However, it also poses challenges related to efficiency, accuracy, security, and fault tolerance. The ongoing research in this domain aims to harness the capabilities of digital twins for creating intelligent and responsive 6G networks that can meet the diverse requirements of emerging applications.

1.2 Basic of Digital Twin

1.2.1 What Are Digital Twin Networks?

A digital twin is an up-to-date representation, a model, of an actual physical asset in operation. It reflects the current operation condition and includes relevant historical data [6]. Digital twin networks can be used to evaluate the current condition of the operation and, more importantly, predict future behavior, refine the control, or optimize the operation. A digital twin can be a model of a component, a system of components, or a system of systems, for example, smart cities, power plants, manufacturing units, or an aircraft. Digital twin models can include physics-based methods or statistical models. The models reflect the operating process's current environment, timeline, and configuration, which typically involves directly streaming operational data into tuning algorithms. In wireless communication, digital twins are crucial in optimizing network performance and ensuring seamless connectivity. These digital replicas of physical assets, such as base stations, antennas, and communication nodes, enable real-time monitoring of network conditions. By integrating data from various sources, such as network infrastructure, user devices, and environmental factors, digital twin networks provide a comprehensive view of the communication ecosystem. One key advantage of digital twins in wireless communication is their ability to predict and mitigate potential network disruptions or congestion. By analyzing historical data and current operational parameters, digital twin models can forecast network load patterns, identify potential bottlenecks, and dynamically adjust resource allocation to optimize performance. This predictive capability is precious in managing the increasing complexity and demand for wireless services in urban environments and densely populated areas. Furthermore, digital twin networks facilitate the rapid deployment and testing of new communication technologies and protocols. Through virtual simulations and emulation of real-world scenarios, stakeholders can evaluate the impact of network upgrades or policy changes before implementation, reducing deployment risks and ensuring smoother transitions. Moreover, digital twins enable iterative refinement of communication algorithms and protocols by providing a platform for continuous testing and optimization, ultimately enhancing wireless networks' reliability, efficiency, and scalability. Digital twins in wireless communication serve as real-time replicas of physical assets, aiding in predictive analysis, optimizing resource allocation, and facilitating the deployment of new technologies, ultimately enhancing network performance and reliability.

1.2.2 How Do Digital Twin Networks Work?

The power of digital twin networks comes from connecting real-world operations or systems with real-world data so we can better visualize them [7]. Digital twins enable cross-functional teams to collaboratively design, build, test, deploy, and operate complex systems in interactive and immersive ways. They help organizations understand the past, view present conditions, and prevent future problems. They inform decision-making through sales and marketing insights, analysis, 3D visualization, simulation, and prediction. In wireless communication, integrating digital twin networks revolutionizes the approach to designing, deploying, and managing complex communication systems. By connecting real-world operations with real-time data streams from network infrastructure, user devices, and environmental sensors, digital twin networks offer a holistic view of the communication ecosystem.

These interconnections enable cross-functional teams, including engineers, network operators, and data analysts, to collaborate effectively in optimizing network performance and addressing emerging challenges. Digital twins facilitate iterative design and testing of wireless communication systems, allowing teams to simulate various deployment scenarios and assess the impact of different configurations on network performance. Through immersive 3D visualization and simulation tools, stakeholders can interact with virtual representations of communication infrastructure, gaining insights into signal propagation, coverage areas, and potential interference sources. This immersive approach enhances the understanding of present network conditions and enables proactive problem-solving and optimization strategies to prevent future disruptions. Moreover, digital twin networks empower organizations to make data-driven decisions by leveraging advanced analytics and predictive modeling. By analyzing historical data and real-time performance metrics, stakeholders can identify trends, forecast demand patterns, and anticipate potential network congestion or failures. These predictive capabilities enable proactive resource allocation, capacity planning, and network optimization, ensuring that wireless communication systems can effectively meet the evolving needs of users while maintaining high reliability and performance levels. These digital twin networks connect real-world operations with real-time data, enabling cross-functional collaboration and proactive problem-solving. Through iterative design and testing, stakeholders can simulate deployment scenarios and optimize network performance. Leveraging advanced analytics and predictive modeling, organizations can make data-driven decisions to anticipate network challenges and ensure reliability and performance.

1.2.3 Why Digital Twin Networks for Wireless Networks?

Digital twin networks use store-and-forward message switching to overcome the problems associated with intermittent connectivity, long or variable delay, asymmetric data rates, and high error rates [8]. In detail, store-and-forward is a net-working technique where data packets are temporarily stored at intermediate points within the network before being forwarded to their destination. This method allows for data transmission in situations where direct end-to-end communication may not be possible due to intermittent connectivity, long delays, or varying data rates. By temporarily storing and then forwarding data packets, store-and-forward mechanisms help overcome challenges associated with unreliable or unpredictable network conditions, ensuring the successful delivery of data even in adverse environments. This approach is commonly used in various network technologies, including satellite communication, where delays and interruptions are more prevalent Sustainable networks are also possible with digital twin networks as they can make networks more resilient. Additionally, by buffering and forwarding messages, digital twin networks can accommodate varying delays and asymmetric data rates, optimizing communication efficiency in dynamic wireless environments. Moreover, digital twin networks enhance the resilience of wireless communication systems, making them more sustainable in the face of environmental uncertainties and operational disruptions. Digital twin networks enable rapid recovery and adaptation to changing conditions by maintaining synchronized digital replicas of network components and operations. Digital twins provide a reference point for restoring operations and reconfiguring network resources in network failures or disturbances, minimizing downtime and service disruptions. Furthermore, leveraging historical data and predictive analytics, digital twin networks can anticipate potential failure scenarios and proactively implement resilience strategies, such as dynamic routing and adaptive resource allocation, to ensure continuous service availability and reliability. Overall, digital twin networks offer a robust framework for building sustainable and resilient wireless communication infrastructures capable of meeting the evolving demands of modern connectivity requirements.

1.3 Architecture of Digital Twin

Digital twin networks enable interactive mapping and control between virtual and natural environments. The recommended digital twin network architecture consists of three layers: the physical, twin, and application layers, as shown in Fig. 1.1.

1.3.1 Physical Network Layer

This layer represents the real-world network, including physical network elements and their operating environment. The physical network layer exchanges data and control messages with the twin layer. The physical network layer employs Various technologies and protocols to enable wireless communication, including radio frequency transmission, modulation techniques, multiple access schemes, and signal propagation mechanisms. These technologies govern how data is transmitted, received, and processed within the wireless communication system, considering signal strength, interference, and channel characteristics. Additionally, the physical network layer interfaces with higher network stack layers, such as the data link layer and network layer, to ensure seamless communication between network elements and end-user devices. Overall, the physical network layer forms the foundation of wireless communication systems, providing the infrastructure and mechanisms necessary for data exchange and transmission in the real-world environment.

1.3.2 Twin Layer

A digital twin network focuses on specific parts of the target network (e.g., radio cell, access network, transport network) or the entire end-to-end network. Data

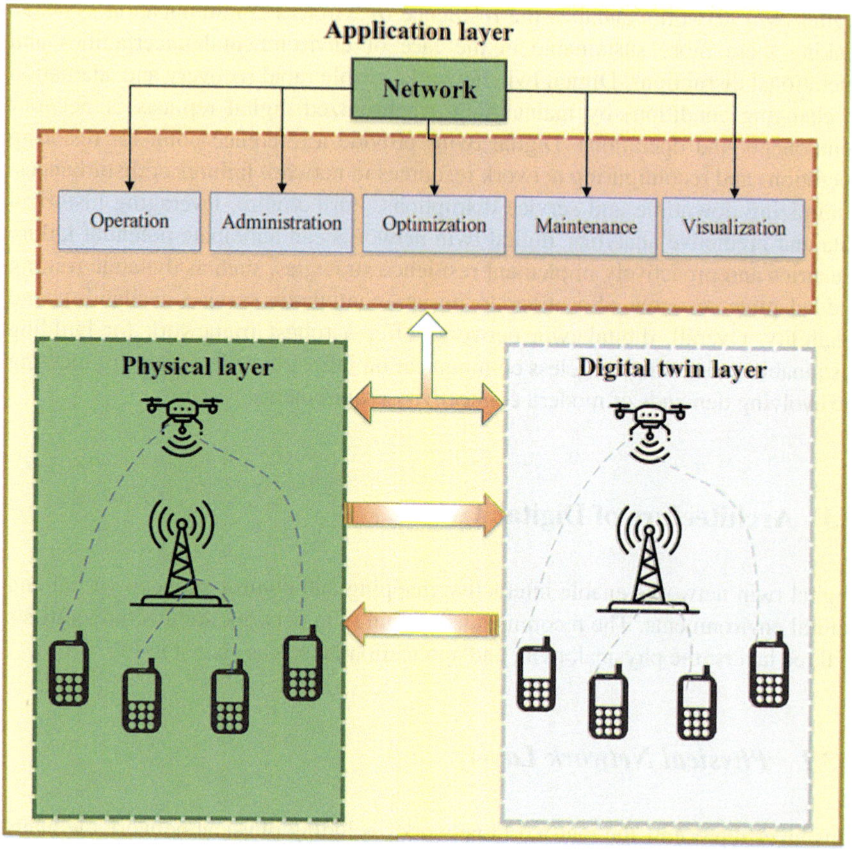

Fig. 1.1 Architecture of digital twin networks

and control messages are exchanged between the physical network layer and the twin layer [9]. The base of a digital twin network comprises three domains: the data domain, the model domain, and the management domain [10]. The data domain collects data from the physical network to build accurate models in the twin. Functions include data collection, storage, service, and management. The model domain is responsible for models representing real-world objects based on collected data. Two models are basic (describing network elements and topology) and functional (analytical models for network planning and traffic analysis). Management domain Manages the twin layer, including model management (creation, configuration, update, monitoring) and security management (authentication, authorization, encryption). Includes various network applications that utilize the twin layer's capabilities. These applications provide intent to the twin layer, which emulates required services and, if necessary, sends control messages to the physical network layer. Examples of network applications include network operations, network administration and maintenance, network optimization, and network visualization [11]. Here, we can highlight the importance of developing standard components, frameworks, standardized interfaces, and universal platforms for digital twin networks. Interoperability and extensibility are crucial to meet diverse use case demands. It suggests reusing existing functionalities from the network operations domain and enhancing them to meet new requirements while simultaneously creating new platforms and capabilities for digital twin networks. The goal is to build digital twin networks with varied capabilities to address various applications effectively.

1.4 Applications of Digital Twins

Now, where will these digital twins work? Everywhere. From ensuring our favorite gadgets are top-notch to optimizing entire industries like healthcare and manufacturing. Digital twin technology offers several applications and benefits in wireless networks, aiding in designing, deploying, managing, and optimizing communication systems. Here are some critical applications, as shown in Fig. 1.2.

1.4.1 Network Planning and Design

Digital twins enable the creation of virtual replicas of wireless networks, allowing for realistic simulations and analysis of different configurations before actual

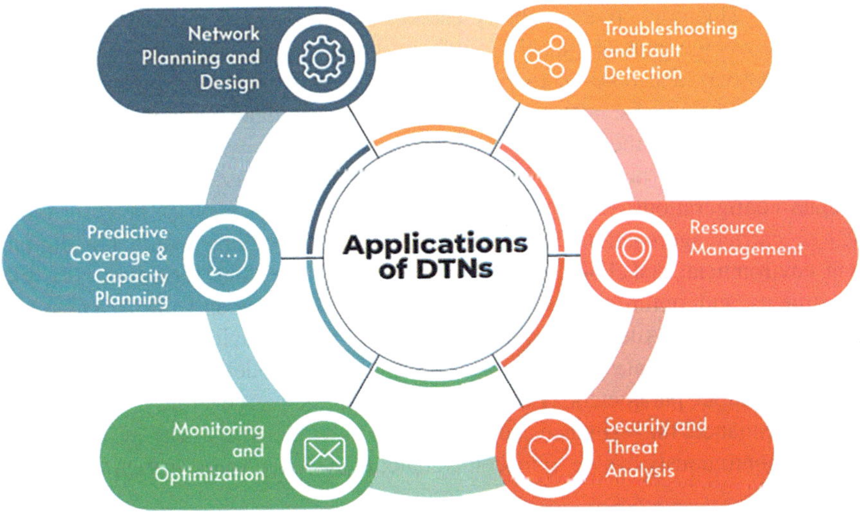

Fig. 1.2 Applications of digital twin networks

deployment. These virtual replicas accurately mirror the physical infrastructure, including base stations, antennas, and propagation characteristics, providing engineers and designers with a comprehensive platform for testing various design alternatives. By simulating different configurations, such as antenna placements, transmission power levels, and network topologies, stakeholders can evaluate the impact of these factors on network performance metrics like coverage, capacity, and signal quality. This enables informed decision-making during the planning phase, ensuring the final network design meets performance objectives and regulatory requirements. Furthermore, digital twins facilitate iterative design processes, allowing for rapid prototyping and optimization of wireless networks. Engineers can quickly modify parameters or experiment with alternative designs within the virtual environment, assessing their impact on real-time performance metrics. This iterative approach identifies optimal design configurations while minimizing the need for costly physical prototypes or field trials. Additionally, digital twins enable collaboration among multidisciplinary teams, including network engineers, radio frequency specialists, and urban planners, fostering creativity and innovation in network design solutions. Overall, digital twins streamline the network planning and design process, accelerating time-to-market and enhancing the efficiency and effectiveness of wireless communication deployments.

1.4.2 Predictive Coverage and Capacity Planning

Digital twins help predict the coverage and capacity of wireless networks based on varying factors such as user density, device types, and environmental conditions. By integrating real-time data streams and historical information into predictive models, digital twins can accurately anticipate future network requirements and performance constraints. This predictive capability enables stakeholders to proactively address potential coverage gaps or capacity bottlenecks before they occur, ensuring seamless connectivity and optimal user experiences. Moreover, digital twins facilitate scenario-based analysis, allowing stakeholders to simulate various deployment scenarios and assess their impact on coverage and capacity. For example, engineers can evaluate the effects of deploying additional base stations, adjusting antenna configurations, or implementing new technologies such as beamforming [12] or carrier aggregation. By conducting these simulations within the digital twin environment, stakeholders can make informed decisions regarding network investments and optimizations, maximizing the efficiency and effectiveness of resource allocation. Additionally, digital twins enable sensitivity analysis, allowing stakeholders to evaluate the robustness of predictive models to uncertainties or changes in input parameters. This approach to predictive coverage and capacity planning enhances the resilience and adaptability of wireless networks, ensuring that they can continuously evolve to meet the changing needs of users and applications.

1.4.3 Monitoring and Optimization

Digital twins facilitate real-time network performance monitoring, including signal strength, latency, and data throughput, enabling quick identification and resolution of issues [13]. Through simulations and analysis, digital twins assist in devising optimization strategies to enhance network performance and resource utilization. In wireless communication, real-time monitoring and optimization are essential for maintaining the reliability and efficiency of networks, especially in dynamic and congested environments. Digital twins are a powerful tool for continuously monitoring network performance metrics such as signal strength, latency, and data throughput. By integrating data from network infrastructure, user devices, and environmental sensors, digital twins provide a comprehensive view of network operations in real time. This real-time monitoring capability enables stakeholders to quickly identify and diagnose performance issues, such as coverage gaps, congestion hotspots, or interference sources, before they escalate into service disruptions [13]. Furthermore, digital twins facilitate proactive optimization strategies by leveraging insights from real-time monitoring and simulation-based analysis. By simulating different scenarios and optimization techniques within the digital twin environment, stakeholders can evaluate the potential impact of proposed changes on network performance and resource utilization [14]. For example, engineers can experiment with dynamic resource allocation algorithms, load balancing strategies, or frequency reassignment schemes to alleviate congestion and improve overall network efficiency. By refining these optimization strategies based on real-time feedback from the digital twin, stakeholders can continuously enhance network performance, scalability, and resilience, ensuring that wireless communication systems can effectively meet the evolving demands of users and applications.

1.4.4 Troubleshooting and Fault Detection

Digital twins can identify abnormal behavior in the network, helping detect faults, security breaches, or performance degradation early. When issues arise, digital twins provide insights into the root causes of problems, supporting more efficient troubleshooting. In wireless communication, troubleshooting and fault detection is paramount for maintaining uninterrupted connectivity and ensuring the reliability of communication services. Digital twins proactively identify abnormal behavior within the network by continuously monitoring key performance indicators and comparing them to predefined thresholds. Through real-time analysis of network data, digital twins can detect deviations from expected norms, such as sudden drops in signal strength, unusually high latency, or unexpected traffic patterns, which may indicate potential faults or security breaches. By flagging these anomalies early, digital twins enable stakeholders to initiate timely troubleshooting measures before they escalate into critical issues, minimizing downtime and user service disruptions.

Moreover, digital twins provide valuable insights into the root causes of network problems, facilitating more efficient troubleshooting and resolution processes. When issues arise, digital twins serve as a virtual environment for diagnosing and analyzing the underlying factors contributing to the problem. By correlating data from various sources, such as network infrastructure logs, performance metrics, and user feedback, digital twins can pinpoint the root causes of faults or performance degradation, whether from hardware failures, software bugs, configuration errors, or external interference. With this contextual information, network operators and engineers can implement targeted remediation strategies, such as software patches, hardware replacements, or configuration adjustments, to address the underlying issues and restore normal network operation swiftly. Digital twins essentially streamline wireless communication systems' troubleshooting and fault detection process by providing real-time monitoring, early anomaly detection, and root cause analysis capabilities. By leveraging insights from digital twins, stakeholders can proactively identify and address network issues, ensuring optimal performance, reliability, and user satisfaction [15]. Additionally, the continuous feedback loop facilitated by digital twins enables organizations to learn from past incidents, refine their troubleshooting procedures, and enhance the overall resilience of wireless communication networks.

1.4.5 Resource Management

Digital twins aid in simulating and optimizing spectrum usage, allowing for efficient allocation of frequencies and mitigating interference. Optimizing network resources, including power consumption of devices and infrastructure, contributes to overall energy efficiency. Digital twins assist in studying the propagation characteristics of millimeter-wave frequencies used in 5G and beyond, considering factors like building structures and atmospheric conditions. Simulating network slicing scenarios helps understand and optimize the deployment of diverse services over a shared infrastructure. Digital twins can model user behavior, predict network usage patterns, and provide insights into improving the overall user experience. Simulating the interactions between various devices within the network helps optimize connectivity and performance. In wireless communication, efficient resource management is essential for maximizing network performance while minimizing costs and environmental impact. Digital twins play a crucial role by enabling simulation and optimization of spectrum usage, which is particularly critical due to the scarcity of available frequency bands. By creating virtual replicas of wireless environments, digital twins allow stakeholders to experiment with different frequency allocations and interference mitigation techniques, predicting their impact on network performance and user experience. This capability enables more effective spectrum planning, ensuring that frequencies are allocated to minimize interference and maximize throughput, ultimately improving the overall efficiency of wireless networks. Additionally, digital twins contribute to energy efficiency by optimizing the power consumption of devices and infrastructure within wireless networks [16]. By modeling the energy consumption profiles of various network components and simulating different operational scenarios, stakeholders can identify opportunities for energy savings and implement strategies to reduce power consumption without compromising performance. This may involve dynamic power management techniques, such as sleep modes for idle devices or adaptive transmission power control, to match power consumption with network demand. By optimizing energy usage across the network, digital twins help reduce operational costs and carbon footprint, contributing to sustainability efforts in the telecommunications industry. Furthermore, digital twins facilitate the deployment and optimization of advanced technologies such as network slicing, which allows multiple virtual networks to coexist on a shared physical infrastructure. By simulating network slicing scenarios within the digital twin environment, stakeholders can evaluate the performance and resource utilization of different slice configurations, optimizing the allocation of network resources to meet the diverse requirements of various services and applications. This ensures efficient utilization of network resources while enabling operators to offer differentiated services tailored to the specific needs of different user groups. Digital twins empower stakeholders in the wireless communication industry to make informed decisions regarding resource management, leading to more efficient, sustainable, and user-centric network deployments.

1.4.6 Security and Threat Analysis

Digital twins enable the simulation of cybersecurity threats and vulnerabilities, allowing for proactive measures to enhance network security. Testing and simulating security policies in a digital twin environment help identify potential weaknesses before implementation. In wireless communication, security is critical due to the susceptibility of wireless networks to various cyber threats and vulnerabilities. Digital twins provide a unique platform for simulating cybersecurity threats and vulnerabilities, enabling stakeholders to assess and strengthen network security defenses proactively. By replicating the network architecture and security protocols within the digital twin environment, stakeholders can simulate various attack scenarios, such as unauthorized access attempts, denial-of-service attacks, or malware propagation, and evaluate the effectiveness of existing security measures in mitigating these threats. This proactive approach to security analysis helps identify potential weaknesses in the network infrastructure or security policies before they are exploited by malicious actors, allowing timely remediation actions to be implemented [17]. Furthermore, digital twins facilitate the testing and validating of security policies and protocols in a controlled environment, reducing the risks associated with deploying untested security measures in production networks. By simulating security policies within the digital twin environment and subjecting them to simulated attack scenarios, stakeholders can assess their efficacy in preventing or mitigating security breaches. This testing process enables stakeholders to fine-tune security policies, adjust configurations, or deploy additional security controls to enhance network resilience and protect against emerging threats. Additionally, by incorporating feedback from security simulations into the design and deployment of wireless networks, stakeholders can ensure that security considerations are integrated into every aspect of network architecture and operations, minimizing the likelihood of security incidents and safeguarding the confidentiality, integrity, and availability of wireless communication services. In summary, digital twins provide a powerful tool for enhancing security and threat analysis in wireless communication networks by enabling proactive simulation of cybersecurity threats, testing security policies, and validating security controls. By leveraging the capabilities of digital twins, stakeholders can strengthen network defenses, mitigate security risks, and protect against evolving cyber threats, ultimately enhancing the trust and reliability of wireless communication services for users.

1.5 Benefits and Challenges

Digital twins are like the superheroes of efficiency, predictive maintenance, and cost reduction. But, we've got challenges to data security and interoperability hurdles that we have to overcome. As a digital twin matures, other benefits include reduced maintenance costs, more informed process change decisions with significant potential savings, and improved maintenance and operational efficiency [18]. Having better designs from the start pays dividends over a project's lifetime, as 80–90% of costs incurred during the production, use, and maintenance of a facility are determined at the design stage. Using digital twins in the design industry has improved multi-user collaboration and communication. Pre-construction clients experience seamless aggregation of data and trade coordination. Every enterprise undergoing a digital transformation risks drowning in raw data before finding a way to process and leverage it. In wireless communication, digital twins offer tremendous benefits in optimizing network performance, enhancing predictive maintenance, and reducing operational costs. By creating virtual replicas of wireless networks, stakeholders can gain insights into network behavior, identify potential issues, and implement proactive maintenance measures to prevent downtime and service disruptions. Additionally, digital twins enable more informed decision-making regarding process changes and optimizations, leading to significant cost savings and operational efficiencies. For instance, by simulating different network configurations and deployment scenarios, stakeholders can identify optimal designs that minimize deployment costs and maximize performance, thus ensuring better utilization of resources over the project's lifetime. However, alongside these benefits, the adoption of digital twins in wireless communication also presents challenges, particularly in data security and interoperability. With the increasing volume and complexity of data generated by wireless networks, ensuring the security and privacy of sensitive information becomes paramount [19]. Stakeholders must implement robust security measures to protect against cyber threats and unauthorized access to digital twin data. Furthermore, interoperability hurdles may arise when integrating digital twins with existing network infrastructure and management systems. Standardization efforts and collaboration among industry stakeholders are essential to address these challenges and establish interoperable frameworks for the seamless integration and operation of digital twins within wireless communication ecosystems. Despite these challenges, as digital twin technology matures, the benefits of wireless communication networks are undeniable. From reduced maintenance costs to improved operational efficiency and enhanced collaboration, digital twins offer a transformative approach to network design, deployment, and management. By leveraging real-time 3D digital twins, stakeholders can unlock the full potential of enterprise and IoT data, enabling better decision-making and optimizing wireless communication systems to benefit users and organizations alike [20]. Today, capturing raw data is less of a challenge than processing it, filtering the useless parts, combining it, and transforming it into information that makes sense to the user in the context of their application. The main challenge is unlocking the power of information. Enterprise and IoT data has been buried in databases, spreadsheets, models, computer-aided design software, geographic information systems, and building information modeling (CAD, BIM, GIS). Real-time 3D digital twins can also bring that data to life in wireless networks.

References

1. F. Karimzadeh, M. Imani, B. Asgari, N. Cao, Y. Lin, Y. Fang, Memory-based computing for energy-efficient AI: Grand challenges, in *2023 IFIP/IEEE 31st International Conference on Very Large Scale Integration (VLSI-SoC)*, (IEEE, 2023), pp. 1–8
2. C. De Alwis, A. Kalla, Q.-V. Pham, P. Kumar, K. Dev, W.-J. Hwang, M. Liyanage, Survey on 6G frontiers: Trends, applications, requirements, technologies and future research. IEEE Open J. Commun. Soc. **2**, 836–886 (2021)
3. S. Mihai, M. Yaqoob, D.V. Hung, W. Davis, P. Towakel, M. Raza, M. Karamanoglu, B. Barn, D. Shetve, R.V. Prasad, et al., Digital twins: A survey on enabling technologies, challenges, trends and future prospects. IEEE Commun. Surv. Tutor. **24**, 2255–2291 (2022)
4. B.R. Barricelli, E. Casiraghi, D. Fogli, A survey on digital twin: Definitions, characteristics, applications, and design implications. IEEE Access **7**, 167653–167671 (2019)
5. M. Ball, *The Metaverse: And How It Will Revolutionize Everything* (Liveright Publishing, 2022)
6. J.M.D. Delgado, L. Oyedele, Digital twins for the built environment: Learning from conceptual and process models in manufacturing. Adv. Eng. Inform. **49**, 101332 (2021)
7. H. Jiang, S. Qin, J. Fu, J. Zhang, G. Ding, How to model and implement connections between physical and virtual models for digital twin application. J. Manuf. Syst. **58**, 36–51 (2021)
8. O. Kodheli, E. Lagunas, N. Maturo, S.K. Sharma, B. Shankar, J.F.M. Montoya, J.C.M. Duncan, D. Spano, S. Chatzinotas, S. Kisseleff, et al., Satellite communications in the new space era: A survey and future challenges. IEEE Commun. Surv. Tutor. **23**(1), 70–109 (2020)
9. C. Gyorgyi, J. Pető, P. Vörös, G. Szabó, S. Laki, Radio propagation digital twin aided multi point transmission with in-network dynamic on-off switching. IEEE Access **11**, 128843–128858 (2023)
10. Y. Wu, K. Zhang, Y. Zhang, Digital twin networks: A survey. IEEE Internet Things J. **8**(18), 13789–13804 (2021)

11. M. Kantor, E. Biernacka, P. Boryło, J. Domżał, P. Jurkiewicz, M. Stypiński, R. Wójcik, A survey on multi-layer IP and optical Software-Defined Networks. Comput. Netw. **162**, 106844 (2019)
12. T.-K. Kim, Effective beamforming technique for carrier aggregation of millimeter wave. Asia Pac. J. Converg. Res. Interchange **5**(1), 21–30 (2019)
13. N. Apostolakis, L.E. Chatzieleftheriou, D. Bega, M. Gramaglia, A. Banchs, Digital twins for next-generation mobile networks: Applications and solutions. IEEE Commun. Mag. **61**, 80–86 (2023)
14. A. Rasheed, O. San, T. Kvamsdal, Digital twin: Values, challenges and enablers from a modeling perspective. IEEE Access **8**, 21980–22012 (2020)
15. K.Y.H. Lim, P. Zheng, C.-H. Chen, A state-of-the-art survey of Digital Twin: Techniques, engineering product lifecycle management and business innovation perspectives. J. Intell. Manuf. **31**, 1313–1337 (2020)
16. D. Xia, J. Shi, K. Wan, J. Wan, M. Martínez-García, X. Guan, Digital twin and artificial intelligence for intelligent planning and energy-efficient deployment of 6G networks in smart factories. IEEE Wirel. Commun. **30**(3), 171–179 (2023)
17. A. Mukherjee, *Network Security Strategies: Protect Your Network and Enterprise Against Advanced Cybersecurity Attacks and Threats* (Packt Publishing Ltd, 2020)
18. T.R. Wanasinghe, L. Wroblewski, B.K. Petersen, R.G. Gosine, L.A. James, O. De Silva, G.K. Mann, P.J. Warrian, Digital twin for the oil and gas industry: Overview, research trends, opportunities, and challenges. IEEE Access **8**, 104175–104197 (2020)
19. W. Lou, Y. Fang, A survey of wireless security in mobile ad hoc networks: Challenges and available solutions, in *Ad Hoc Wireless Networking*, (Springer, 2003), pp. 319–364
20. N. Tripathi, H. Hietala, Y. Xu, R. Liyanage, Stakeholders collaborations, challenges and emerging concepts in digital twin ecosystems. Inf. Softw. Technol. **169**, 107424 (2024)

Chapter 2
Dielectric and Path Loss Modeling to Support Simple and Fast Digital Twins for Wireless Human Body Area Network

Kok Yeow You, Rozeha A. Rashid, Nor Aishah Muhammad, Omar Abdul Aziz, Mohd Adib Sarijari, Ahmad Shahidan Abdullah, and Samura Ali

Abbreviations

$CaCl_2.2H_2O$	Calcium Chloride Dihydrate
CENELEC	European Committee for Electrotechnical Standardization
CT	Computed Tomography
DC	Direct Current
DGBE	Diethylene Glycol Butyl Ether
DT	Digital Twin
FCC	Federal Communications Commission
FD	Finite-Difference
FDTD	Finite-Difference Time-Domain
FE	Finite-Element
FV	Finite-Volume
GUI	Graphical User Interface
HCl	Hydrochloric
HEC	Hydroxyethyl Cellulose
IEEE	Institute of Electrical and Electronics Engineers
IoBT	Internet of BioThings
IoT	Internet of Things
ISM	Industrial, Scientific, and Medical
ITIS	Foundation for Research on Information Technologies in Society
IU	User Interface
KCl	Potassium Chloride

K. Y. You (✉) · R. A. Rashid · N. A. Muhammad · O. A. Aziz · M. A. Sarijari · A. S. Abdullah · S. Ali
Faculty of Electrical Engineering, Universiti Teknologi Malaysia, Skudai, Johor, Malaysia
e-mail: kyyou@fke.utm.my

M. K. Afzal et al. (eds.), *Digital Twins for Wireless Networks*,
https://doi.org/10.1007/978-3-031-73679-7_2

LoS	Line-of-Sight
MATLAB	Matrix Laboratory
$MgCl_2 \cdot 6H_2O$	Magnesium Chloride Hexahydrate
MRI	Magnetic Resonance Imaging
NaCl	Sodium Chloride
$NaHCO_3$	Sodium Hydrogen Carbonate
$Na_2HPO_4.2H_2O$	Sodium Dihydrogen Phosphate Dihydrate
Na_2SO_4	Sodium Sulphate
NLoS	Non-Line-of-Sight
PDE	Partial Differential Equation
PET	Positron Emission Tomography
RF	Radio-Frequency
RMSE	Root Mean Squared Estimator
SAR	Specific Absorption Rate
SHBP	Synthetic Human Blood Plasma
TX-100	Triton X-100
UWB	Ultra-Wideband
WBAN	Wireless Body Area Network
$(CH_2OH)_3CNH_2$	Tris (Hydroxymethyl) Aminometane

2.1 Introduction

Recently, a lot of emerging digital twins (DT) and microwave electronic technologies have been implemented in the biomedical field to improve healthcare service quality, such as the Internet of BioThings (IoBT), big data analytics or artificial intelligence in healthcare, and Wireless Human Body Area Network (WBAN) [1, 2]. This chapter only focuses on the TD of WBAN, which is divided into three subtopics, namely, dielectric modeling of body tissues/organs, synthetic fluids simulating actual body tissues/organs, and path loss modeling and simulation in the human body.

In the early stages of biomedical wireless device design, the relative complex permittivity, ε_r value of human tissue play an important role. The interaction between the human body with microwave wireless signal is linked by the relative complex permittivity, ε_r $(= \varepsilon_r' - j\varepsilon_r'')$ [3–6]. However, each type of tissue in the human body has a specific permittivity value. Hence, the path loss of the wireless signals in different parts of the human body is different and makes the study of path loss wireless signals in the human body more complicated. Typically, the complex permittivity, ε_r value for each type of tissue is obtained from *in vivo* or *in vitro* real human body measurements. In addition, human body tissues, such as the heart, vein, and brain can be considered as polar semi-liquid specimens and their measured permittivity values can be modeled by Debye or Cole-Cole models [2, 3]. In this study, the integration between Cole-Cole models for various kinds of human body tissues/organs and MATLAB-based graphical user interface (GUI) is created for human body dielectric DT.

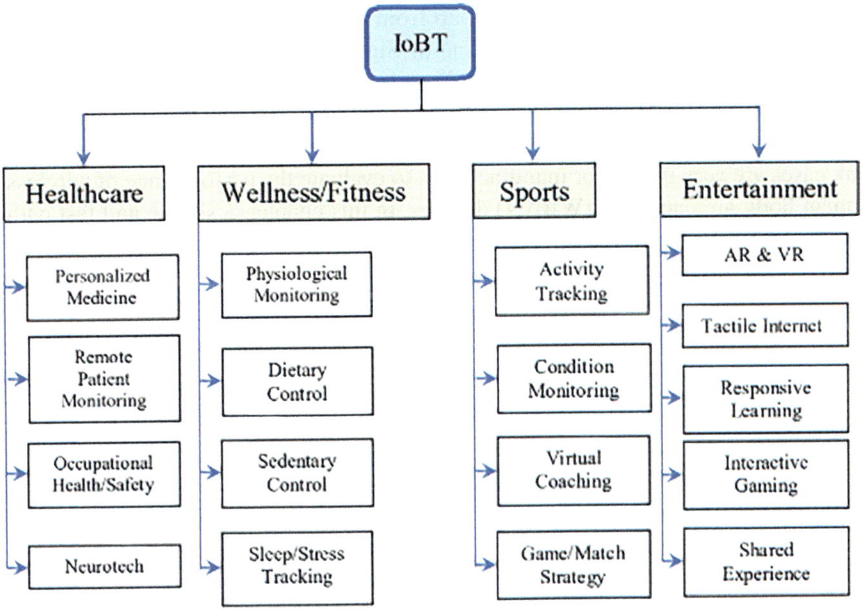

Fig. 2.1 Taxonomy of IoB applications [1]

In the actual test, calibration, and measurement of the performance of wireless devices, it is necessary to develop artificial human tissue. Therefore, various mixed formulations of chemical components have previously been proposed to prepare artificial tissues with electrical characteristics equivalent to actual human tissues. In general, the chemicals used to create artificial human tissue are nontoxic and mild [7–12]. Adjust the mixing ratio of various chemical reagents so that the permittivity value is the same as the measured permittivity of real tissue [6]. In this chapter, some recipes for preparing artificial body tissues, such as human head fluid, breast tissues, and blood plasma, using chemical reagents are introduced. The recipes of the synthetic body tissues are collected and built-in in MATLAB-based GUI to develop synthetic body tissues TD.

There are two main components of the WBAN, consisting of the Internet of BioThings (IoBT) (or sensors) and wireless communication, such as Bluetooth, Zigbee, and Wi-Fi. The IoBT is utilized for monitoring vital signs, including body temperature, blood pressure, glucose levels, and more, as depicted in Fig. 2.1 [1]. Recently, a wireless human body area network (WBAN) has been implemented referring to the IEEE 802.15.6-2012 standard for short-range, low-power, and highly reliable wireless communication around the human body [13]. There are four main possible communication links between IoBTs of the WBAN, namely (1) communication between an implanted IoBT and another implanted IoBT, (2) communication between an implanted IoBT and other IoBTs on the body surface, (3) communication between two IoBTs on the human body surface, and (4) IoBT on the surface body to IoT in free space [14]. The performance and efficiency of wireless devices, such as implanted IoBT used in healthcare services depend largely on the path loss signals in the human body. Apart from the signal transmitted distance, *d* in both Line-of-Sight (LOS) and Non-Line-of-Sight (NLOS) scenarios, the wireless signal inside the human body suffers from losses which can also be interpreted in terms of specific absorption rate (SAR) [15]. Recently, various semi-empirical path loss models have been developed. Path loss models for various communication link cases are very useful for manufacturers to evaluate the performance of wireless human body area network (WBAN) devices. In this chapter, a simple and fast path loss simulation for TD is developed aided with MATLAB GUI [16].

2.1.1 Dielectric Properties of Human Tissues/Organs

The interaction between human body tissues/organs containing water with the wireless signal can be evaluated by the parameter of the relative complex permittivity, ε_r. The relative complex permittivity, ε_r ($= \varepsilon_r' - j\varepsilon_r''$) is typically represented by two components, namely ε_r' (real part) and ε_r'' (imaginary part), where ε_r' represents the ability to store energy, and ε_r'' represents energy loss as heat. However, human tissues and organs exhibit different ε_r values due to variations in their composition, water content, and cellular structure.

In fact, the ε_r plays a critical role in medical applications, particularly in the field of medical imaging. For instance, magnetic resonance imaging (MRI) or microwave imaging relies on the different ε_r of tissues to create detailed images as shown in Fig. 2.2 [17]. In radio-frequency (RF) ablation, ε_r helps determine the effectiveness

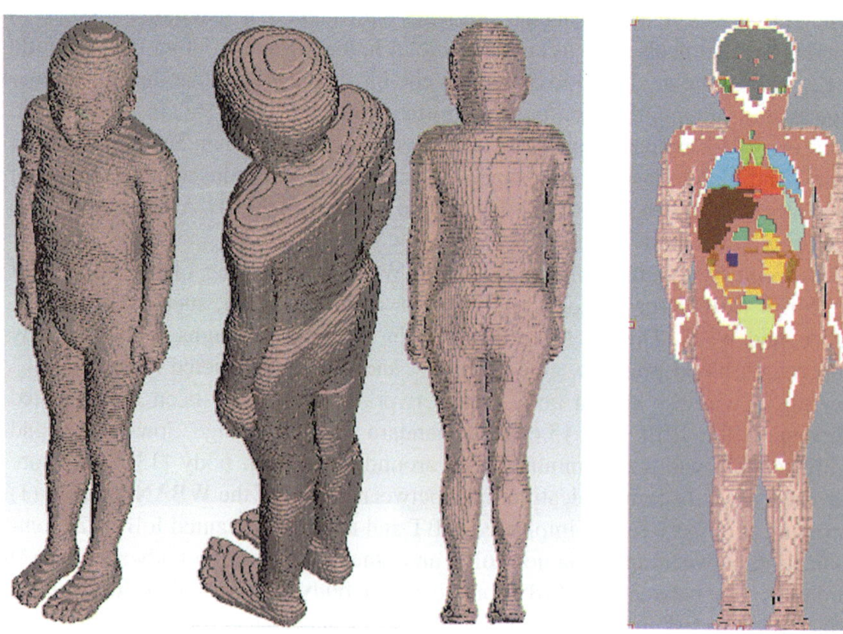

Fig. 2.2 Child body imaging [17]

of using RF energy to treat tumors or abnormal tissue, as well as aids in the design of medical devices, such as pacemakers or wireless implants. It ensures that these devices operate safely within the body without causing harm due to excessive heat or interference. Tissues with higher ε_r may heat up more efficiently.

2.1.2 Debye/Cole-Cole Models

Accurate dielectric property data for human tissues are essential for realistic simulations of electromagnetic field interactions, informing the use of radiation in medical and communication technologies. Most of the human body tissues is a polar semi-solid specimens whose relative complex permittivity, ε_r can be mathematically modeled as [3]:

$$\begin{aligned} \varepsilon_r &= \varepsilon_\infty + \sum_{n=1}^{N} \frac{\Delta_n}{1+(jf/f_{rn})^{1-\alpha_n}} - \frac{j\sigma}{2\pi\varepsilon_o f} \\ &= \varepsilon_r' - j\varepsilon_r'' \end{aligned} \tag{2.1}$$

where ε_r', ε_r'', and σ (in S/m) are the dielectric constant, loss factor, and DC conductivity of the tissues, respectively. Symbol Δ is represented as $\Delta = \varepsilon_s - \varepsilon_\infty$ which ε_s and ε_∞ are the static and infinite dielectric constant, respectively. On the other hand, ω (in rad/s) is the angular frequency, τ (in ps) is the relaxation time constant, α is the exponent parameter ($0 \le \alpha \le 1$) and ε_o ($\approx 8.8542 \times 10^{-12}$ F/m) is the permittivity of the free space.

To commence our discussion in this section, we will employ a Debye/Cole-Cole model on human breast tissue due to breast tissue encompasses diverse groups with a broad spectrum of permittivity values. Healthy breast tissue is intricate and comprises adipose, glandular, and fibronectin tissues with different fraction percentages of the component tissues in the normal breast as shown in Fig. 2.3

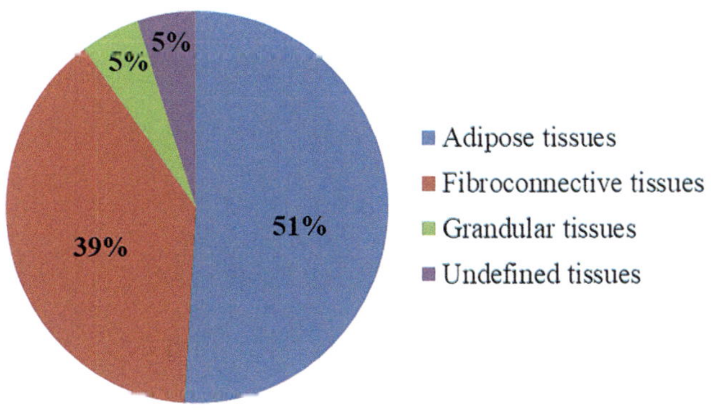

Fig. 2.3 Distribution of component tissues in the normal breast

[4]. Hence, three groups of healthy breast samples with different fraction content of the adipose are analyzed, namely Group 1 sample containing 0–30% adipose tissues, Group 2 with 31–84% adipose tissues, and Group 3 with 85–100% adipose tissues [4]. Group 3 samples represented the tissues which dominated by the adipose component with very low water content. While, Group 1 and Group 2 samples are the bulk tissues dominated by the glandular and fibroconnective components, respectively that contain higher water content.

Romeo et al. [11] introduced a single-relaxation Cole-Cole model with DC conductivity term, σ to represent relative complex permittivity, ε_r in different groups of breast tissue up to 6 GHz as:

$$\varepsilon_r = \varepsilon_\infty + \frac{\Delta}{1 + (j2\pi f\tau)^{1-\alpha}} - \frac{j\sigma}{2\pi\varepsilon_o f} \tag{2.2}$$

The Cole-Cole parameters in Eq. (2.2) are tabulated in Table 2.1. Figure 2.4 shows that the relative dielectric constant, ε_r', and loss factor, ε_r'' of the 0–30%, 31–84%, and 85–100% adipose content normal breast tissues and tumor breast tissues are

Table 2.1 Cole-Cole parameters value (up to 6 GHz) for normal and tumour breast samples from tumour surgeries

Breast tissues	Tissue group	Percentile (%)	Cole-Cole parameters				
			ε_∞	Δ	$\tau \times 10^{-12}$ (s)	α	σ (S/m)
Normal tissues	Group 1	25	5.013	40.60	10.16	0.091	0.607
	Group 2 + Group 3	75					
	Group 1	50	7.237	46.00	10.30	0.049	0.808
	Group 2 + Group 3	50					
	Group 1	75	7.816	50.21	10.47	0.055	0.889
	Group 2 + Group 3	25					
	Group 2	25	3.891	4.113	13.83	0.038	0.082
	Group 1 + Group 3	75					
	Group 2	50	6.080	19.26	11.47	0.057	0.297
	Group 1 + Group 3	50					
	Group 2	75	6.381	32.30	10.41	0.081	0.561
	Group 1 + Group 3	25					
	Group 3	25	3.122	2.133	14.27	0.099	0.034
	Group 1 + Group 2	75					
	Group 3	50	3.581	3.337	15.21	0.052	0.053
	Group 1 + Group 2	50					
	Group 3	75	3.882	5.020	12.92	0.059	0.103
	Group 1 + Group 2	25					
Tumour	Group 1	>75	9.058	51.31	10.84	0.022	0.899
	Group 2 + Group 3	<25					

Note: Group 1: 0–30% of adipose (70–100% of water); Group 2: 31–84% of adipose (16–69% of water); Group 3: 85–100% of adipose (0–15% of water)

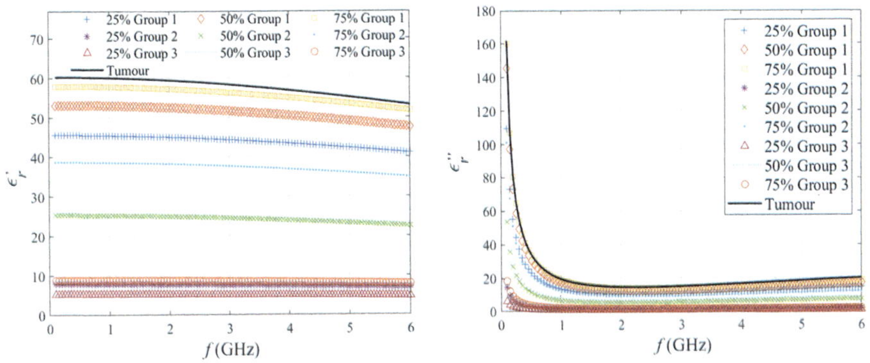

Fig. 2.4 Relative complex permittivity, ε_r $(= \varepsilon_r' - j\varepsilon_r'')$ values of various breast tissue groups up to 6 GHz

significantly distinct. First, the adipose and glandular tissue contents of the cancer samples are very low. Specifically, 50 of the 60 cancer samples (83%) contained 0–20% adipose tissue, and all 60 cancer samples contained 10% or less glandular tissue. The dielectric constant, ε_r of tumor tissue is above the value of 50, it is higher than normal tissues, since the adipose (Group 3) and glandular (Group 2) tissue contents of the cancer tissues are very low, namely cancer tissues typically contained 0–20% adipose tissue and contains 10% or less glandular tissue.

Moreover, a digital twin (DT) platform for predicting the relative complex permittivity, ε_r of breast tissue is developed, utilizing a MATLAB-based graphical user interface (GUI). The MATLAB-based DT is developed using the "`appdesigner`" environment (MATLAB) [16] as shown in Fig. 2.5.

Besides human breast tissue, Kensuke et al. [3] employed a two-relaxation Cole-Cole model to characterize 43 types of tissues/organs, as depicted in formula (2.3). The parameters defined in Eq. (2.3) can be found in Table 2.2.

$$\varepsilon_r = \varepsilon_\infty + \frac{\Delta_1}{1 + \left(jf/f_{r1}\right)^{1-\alpha_1}} + \frac{\Delta_2}{1 + \left(jf/f_{r2}\right)^{1-\alpha_2}} - \frac{j\sigma}{2\pi\varepsilon_o f} \tag{2.3}$$

Likewise, Eq. (2.3) with the parameter values from Table 2.2 is implemented within the MATLAB GUI, as shown in Fig. 2.6.

2.1.3 Permittivity Modeling of Human Body Tissues/Organs

In this section, the steps to determine the Debye/Cole-Cole model's parameters for tissue/organ using an explicit method [18] are demonstrated. This study employs a comprehensive Debye model with three relaxation terms to model the relative complex permittivity, ε_r for an additional 48 types of tissues/organs (those not listed in Table 2.2) across the frequency range from 100 MHz to 20 GHz. The three

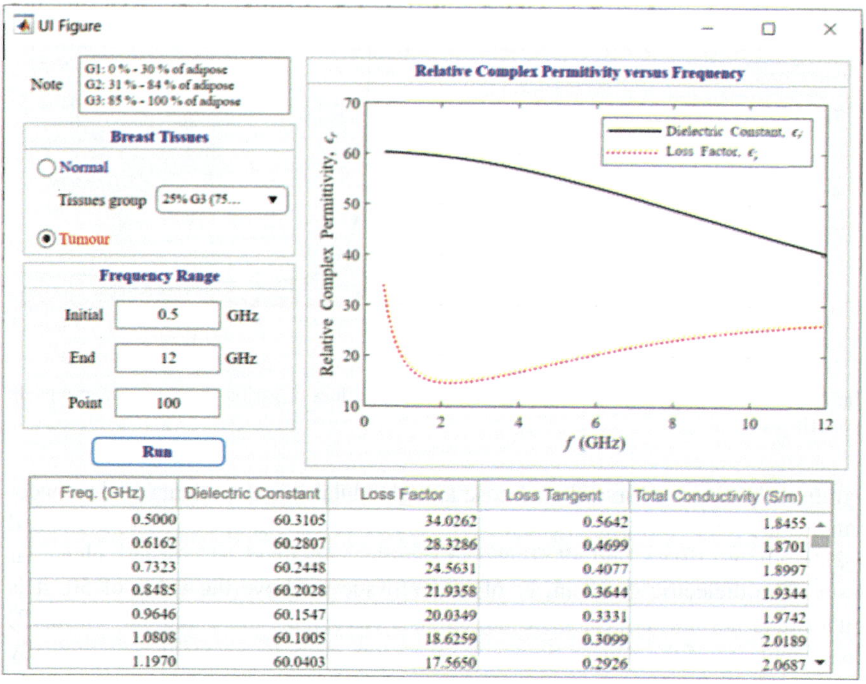

Fig. 2.5 MATLAB-based user interface (IU) of the breast tissue dielectric calculator

relaxation models are given as:

$$\varepsilon_r = \varepsilon_\infty + \frac{\Delta_1}{1 + (jf/f_{r1})} + \frac{\Delta_2}{1 + (jf/f_{r2})} + \frac{\Delta_3}{1 + (jf/f_{r3})} - \frac{j\sigma}{2\pi\varepsilon_o f} - j\zeta \qquad (2.4)$$

where ζ is the error term of the imaginary part at high operating frequency which is represented in linear expression (see Table 2.3). This dielectric modeling is founded on permittivity data acquired from the Foundation for Research on Information Technologies in Society [6]. The modeling process involves several key steps:

Step 1:

The Δ_1, Δ_2, Δ_3, f_{r_1}, f_{r_2}, and f_{r_3} are found from plotting data, namely ε_r' versus $-\omega\varepsilon_r''$ plot. The relaxation time, τ is the gradient of the linear line of the ε_r' versus $-\omega\varepsilon_r''$ plot [18]. In this modeling, we'll use adrenal gland tissue as an example, and it's demonstrated that up to three primary relaxation times occur within the frequency range of 100 MHz to 20 GHz (refer to Fig. 2.7). The relationship between τ and f_r is given as:

$$f_{r1} = \frac{1}{2\pi\tau_1}, \quad f_{r2} = \frac{1}{2\pi\tau_2}, \quad \text{and} \quad f_{r3} = \frac{1}{2\pi\tau_3}$$

Table 2.2 Cole-Cole parameters value (up to 20 GHz) for human body tissues/organs [3]

No	Tissues/organs	Cole-Cole parameters							
		ε_∞	Δ_1	Δ_2	$f_{r\_1} \times 10^9$ (Hz)	$f_{r\_2} \times 10^6$ (Hz)	α_1	α_2	σ (S/m)
1	Aorta	2.27	28.2	1360	15.2	0.0225	0.001	0.590	0.292
2	Bladder	4.37	13.2	1350	12.6	0.846	0.131	0.077	0.178
3	Blood	11.8	41.4	7930	15.8	0.947	0.015	0.092	0.996
4	Bone cancellous	4.21	12.1	917	15.6	0.471	0.003	0.330	0.0231
5	Bone cortical	3.22	10.0	693	9.49	0.414	0.218	0.217	0.018
6	Bone marrow	2.37	2.30	64.6	18.0	1.69	0.002	0.346	0.0022
7	Breast fat	2.11	2.94	14.1	15.3	6.28	0.140	0.401	0.026
8	Cartilage	4.34	35.6	17,600	13.6	0.175	0.257	0.141	0.070
9	Cerebellum	4.41	38.3	1740	20.1	3.22	0.008	0.212	0.105
10	Cervix	3.57	45.0	337	15.5	6.83	0.176	0.263	0.554
11	Cerebrospinal fluid	19.4	47.8	138,000	16.6	0.252	0.029	0.012	0.156
12	Colon	4.56	51.5	2720	17.6	1.80	0.036	0.158	0.309
13	Cornea	3.62	48.4	13,000	18.6	0.845	0.089	0.065	0.227
14	Dura	2.35	43.3	39.7	18.5	74.8	0.082	0.302	0.579
15	Eye sclera	3.42	49.9	9400	17.0	0.661	0.091	0.085	0.377
16	Fat	2.32	2.14	19.7	20.1	8.25	0.002	0.259	0.0222
17	Gall bladder	4.56	48.6	53.5	19.2	96.4	0.022	0.350	0.822
18	Gall bladder bile	2.07	66.3	67.1	23.8	64.6	0.121	0.195	1.29
19	Grey matter	12.7	32.2	1170	18.7	2.36	0.075	0.315	0.173
20	Heart muscle	3.84	49.1	2520	18.7	1.73	0.049	0.196	0.310
21	Kidney	4.43	43.3	2050	20.2	2.87	0.052	0.212	0.267
22	Lens cortex	2.13	48.0	3220	18.1	1.26	0.095	0.118	0.234
23	Lens nucleus	4.62	32.0	2640	14.5	0.368	0.166	0.190	0.137
24	Liver	4.52	36.8	2550	16.9	1.12	0.107	0.219	0.140
25	Lung (deflated)	2.07	47.8	985	17.3	2.71	0.130	0.132	0.349

(continued)

Table 2.2 (continued)

No	Tissues/organs	Cole-Cole parameters							
		ε_∞	Δ_1	Δ_2	$f_{r\_1} \times 10^9$ (Hz)	$f_{r\_2} \times 10^6$ (Hz)	α_1	α_2	σ (S/m)
26	Lung (inflated)	4.44	14.6	646	16.1	3.22	0.001	0.181	0.0842
27	Muscle (parallel)	3.91	55.0	663	18.0	0.78	0.165	0.387	0.633
28	Muscle (transverse)	4.32	59.3	23,500	15.3	0.175	0.240	0.093	0.354
29	Nerve	5.12	24.7	1580	20.9	1.19	0.053	0.251	0.0789
30	Ovary	2.47	37.2	1040	15.1	2.61	0.077	0.312	0.359
31	Skin (dry)	5.01	31.4	830	19.9	6.77	0.066	0.139	0.005
32	Skin (wet)	2.64	45.7	3660	17.6	1.47	0.177	0.097	0.130
33	Small intestine	4.58	52.6	6620	15.5	1.42	0.042	0.160	0.820
34	Spleen	3.40	49.4	2180	19.5	3.43	0.080	0.164	0.138
35	Stomach	2.56	69.2	1400	18.2	3.06	0.051	0.122	0.662
36	Tendon	6.81	28.8	570	14.0	0.0781	0.007	0.607	0.428
37	Testis	3.75	55.8	9410	17.5	0.507	0.051	0.141	0.121
38	Thyroid	5.22	54.1	20,000	15.0	0.216	0.103	0.104	0.230
39	Tongue	3.83	53.0	6000	13.5	0.600	0.149	0.112	0.282
40	Trachea	3.29	41.3	5400	18.1	0.138	0.005	0.255	0.331
41	Uterus	3.28	56.5	1360	18.7	3.39	0.108	0.193	0.464
42	Vitreous humor	2.55	64.9	7000	20.8	0.0425	0.00	0.069	1.48
43	White matter	17.4	15.0	875	9.46	1.52	0.00	0.343	0.0908

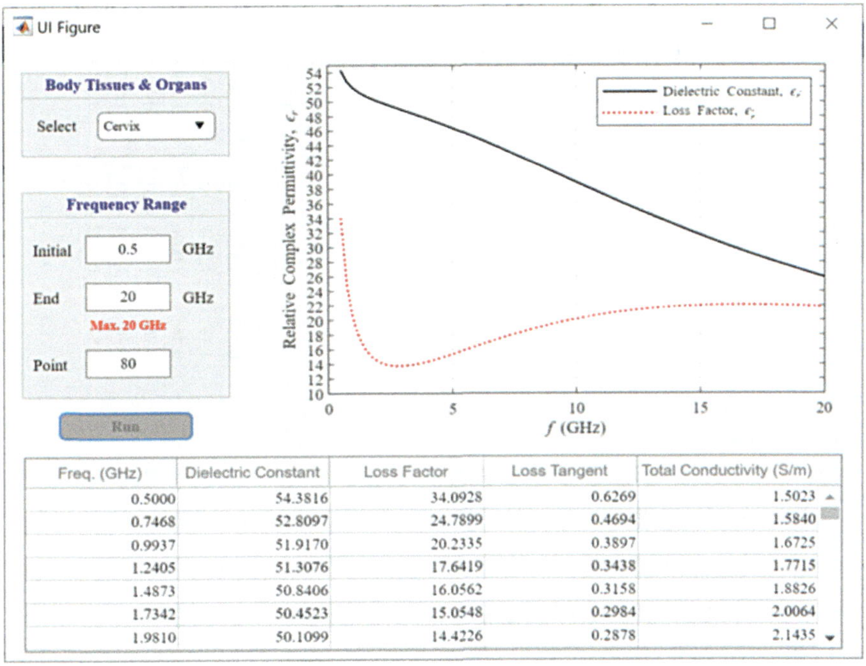

Fig. 2.6 MATLAB-based user interface (IU) of the 43 types of human tissue dielectric calculator

The value of Δ represents the range between the maximum and minimum value of ε_r' across the frequency span for each relaxation, as illustrated in Fig. 2.7. For instance, the values Δ_1, Δ_2, and Δ_3 of the adrenal gland tissue are determined for the frequency ranges of 100 MHz to 400 MHz, 400 MHz to 3.5 GHz, and 3.5 GHz to 20 GHz, respectively.

Step 2:
From step 1, the values $\Delta_1 = 3.93$, $\Delta_2 = 3.6$, $\Delta_3 = 2.4$, $f_{r\_1} = 0.69$ GHz, $f_{r\_2} = 2.7$ GHz, and $f_{r\_3} = 16.48$ GHz are obtained. In an ideal scenario, the value of ε_∞ remains constant for tissues with single relaxation characteristics, determined at high operating frequencies ($f > 300$ GHz). However, many tissues exhibit multi-relaxation properties, including adrenal gland tissue. Additionally, the study modeling is constrained to a maximum operating frequency of 20 GHz. Therefore, in this context, the value of ε_∞ becomes frequency-dependent and is no longer constant.

In Step 2, we incorporate all values of Δ_1, Δ_2, Δ_3, f_{r_1}, f_{r_2}, and f_{r_3} into the three relaxation components of the Debye model. The variation between the ε_r' data and the Debye model's relaxation segments yields the value of ε_∞ as:

$$\varepsilon_\infty = \varepsilon_r' - \left\{ \frac{3.93}{1 + \left(j \frac{f}{0.68 \times 10^9} \right)} + \frac{3.6}{1 + \left(j \frac{f}{2.7 \times 10^9} \right)} + \frac{2.4}{1 + \left(j \frac{f}{16.48 \times 10^9} \right)} \right\} \tag{2.5}$$

Table 2.3 Fitting Cole-Cole parameters value (100 MHz to 20 GHz) for human body tissues/organs based on data from [6]

No	Tissues/organs	Cole-Cole parameters								
		ε_∞	Δ_1	Δ_2	Δ_3	$f_{r\_1} \times 10^9$ (Hz)	$f_{r\_2} \times 10^9$ (Hz)	$f_{r\_3} \times 10^9$ (Hz)	σ (S/m)	High-frequencies loss, ζ
1	Adrenal gland	$-5.4983 \times 10^{-11}f + 53.5$	3.93	3.6	2.4	0.685	2.696	16.48	0.63	$1.5458 \times 10^{-5}f^{0.5} - 0.4735$
2	Bile	$-7.6328 \times 10^{-10}f + 38.5$	29.3	4.3	28.3	0.11	10.3	20.1	1.44	$1.1366 \times 10^{-4}f^{0.5} - 0.55724$
3	Brain	$-2.6256 \times 10^{-10}f + 21.7$	59.4	10.3	15.2	0.15	7.1	18.7	0.64	$6.3738 \times 10^{-5}f^{0.5} - 1.9352$
4	Breast gland	$-5.5577 \times 10^{-10}f + 30.9$	11.2	4.8	23.6	0.18	7.64	18.4	0.76	$7.9936 \times 10^{-5}f^{0.5} - 0.30713$
5	Bronchi	$-3.3712 \times 10^{-10}f + 20.5$	12.3	5.0	16.0	0.20	6.64	18.4	0.51	$5.6176 \times 10^{-5}f^{0.5} - 0.59611$
6	Commissura anterior	$-2.6052 \times 10^{-10}f + 18.7$	23.1	5.1	14.7	0.16	4.71	18.1	0.26	$4.8767 \times 10^{-5}f^{0.5} - 0.93517$
7	Connective tissue	$-3.0039 \times 10^{-10}f + 16.1$	12.2	5.2	23.8	0.14	6.15	13.2	0.45	$2.9102 \times 10^{-5}f^{0.5} + 0.75353$
8	Diaphragm	$-4.9087 \times 10^{-10}f + 30.3$	14.7	4.7	20.0	0.14	6.69	19.7	0.66	$7.9056 \times 10^{-5}f^{0.5} - 0.65165$
9	Ductus deferens	$-3.6682 \times 10^{-10}f + 21.2$	19.5	4.4	18.7	0.12	6.64	17.0	0.40	$4.9668 \times 10^{-5}f^{0.5} + 0.031037$
10	Epididymis	$-5.0747 \times 10^{-10}f + 30.0$	19.0	5.7	24.2	0.18	6.54	18.3	0.86	$8.1836 \times 10^{-5}f^{0.5} - 0.91936$
11	Esophagus	$-5.9321 \times 10^{-10}f + 33.2$	16.3	5.6	25.7	0.18	7.14	18.4	0.85	$8.9774 \times 10^{-5}f^{0.5} - 0.66791$
12	Extracellular fluids	$-6.8466 \times 10^{-10}f + 36.2$	24.6	4.9	27.5	0.13	8.5	18.5	2.03	$9.1401 \times 10^{-5}f^{0.5} - 0.1863$
13	Eye (aqueous humor)	$-6.8466 \times 10^{-10}f + 36.2$	24.6	4.9	27.5	0.13	8.5	18.5	2.03	$9.1401 \times 10^{-5}f^{0.5} - 0.1863$
14	Eye (choroid)	$-5.3233 \times 10^{-10}f + 29.8$	19.8	6.0	24.9	0.17	6.58	17.7	1.17	$7.8797 \times 10^{-5}f^{0.5} - 0.46714$
15	Eye (ciliary body)	$-4.983 \times 10^{-10}f + 30.4$	13.5	4.4	20.0	0.19	7.1	19.7	0.67	$8.2815 \times 10^{-5}f^{0.5} - 1.0567$
16	Eye (retina)	$-3.6539 \times 10^{-10}f + 25.0$	37.4	7.0	19.3	0.18	5.82	18.2	0.45	$7.3337 \times 10^{-5}f^{0.5} - 1.6686$
17	Eye (vitreous humor)	3.8	65.7	0	0	22.0	0	0	1.49	0
18	Fat (not infiltrated)	$-2.386 \times 10^{-11}f + 3.94$	0.8	0.37	1.1	0.19	5.9	17.5	0.034	$3.3581 \times 10^{-6}f^{0.5} - 0.0058795$
19	Heart lumen	$-5.0943 \times 10^{-10}f + 29.1$	19.8	6.0	25.5	0.18	6.6	17.7	1.18	$7.5847 \times 10^{-5}f^{0.5} - 0.80691$
20	Hippocampus	$-3.936 \times 10^{-10}f + 25.2$	35.8	6.3	20.4	0.16	4.7	18.1	0.46	$7.0076 \times 10^{-5}f^{0.5} - 1.1996$
21	Hypophysis	$-5.3935 \times 10^{-10}f + 30.4$	11.1	4.8	24.0	0.20	7.6	18.4	0.76	$7.9361 \times 10^{-5}f^{0.5} - 0.4332$
22	Hypothalamus	$-3.5457 \times 10^{-10}f + 24.6$	35.6	7.3	20.4	0.16	4.6	18.0	0.46	$7.1461 \times 10^{-5}f^{0.5} - 1.6675$
23	Intervertebral disc	$-5.5012 \times 10^{-11}f + 29.2$	8.2	5.7	6.3	0.64	3.6	16.6	0.84	$1.7442 \times 10^{-5}f^{0.5} - 0.3819$

24	Large intestine	$-4.5443 \times 10^{-10}f + 27.7$	29.1	6.6	22.7	0.19	4.9	18.1	0.61	$7.8169 \times 10^{-5}f^{0.5} - 1.3394$
25	Large intestine lumen	$-4.9525 \times 10^{-10}f + 30.3$	13.6	4.4	20.1	0.17	7.1	19.7	0.67	$8.097 \times 10^{-5}f^{0.5} - 0.86912$
26	Larynx	$-3.0218 \times 10^{-10}f + 16.8$	16.6	7.5	18.2	0.18	3.5	12.8	0.43	$3.6305 \times 10^{-5}f^{0.5} + 0.013586$
27	Lymphnode	$-1.8166 \times 10^{-11}f + 73.5$	3.6	3.3	2.1	0.49	4.1	19.1	0.59	$6.1178 \times 10^{-6}f^{0.5} + 0.049463$
28	Mandible	$-2.152 \times 10^{-11}f + 4.96$	2.7	2.2	5.1	0.30	3.7	12.5	0.059	$6.6818 \times 10^{-6}f^{0.5} - 0.20179$
29	Medulla oblongata	$-3.1781 \times 10^{-10}f + 22.7$	56.1	7.8	17.5	0.15	4.7	18.0	0.63	$6.5647 \times 10^{-5}f^{0.5} - 1.5623$
30	Meniscus	$-1.8553 \times 10^{-10}f + 14.3$	16.6	7.9	19.2	0.22	4.7	12.7	0.43	$2.8719 \times 10^{-5}f^{0.5} - 0.23059$
31	Midbrain	$-3.6585 \times 10^{-10}f + 23.6$	53.6	7.1	18.0	0.15	3.6	17.9	0.64	$6.8737 \times 10^{-5}f^{0.5} - 1.515$
32	Mucous membrane	$-4.9067 \times 10^{-10}f + 30.2$	13.5	4.5	20.1	0.18	7.1	19.7	0.67	$8.1504 \times 10^{-5}f^{0.5} - 0.98095$
33	Pancreas	$-5.451 \times 10^{-10}f + 30.6$	11.1	4.9	23.8	0.20	7.6	18.4	0.76	$8.0107 \times 10^{-5}f^{0.5} - 0.47121$
34	Penis	$-3.6517 \times 10^{-10}f + 21.2$	19.6	4.5	18.6	0.15	6.6	17.0	0.41	$5.3562 \times 10^{-5}f^{0.5} - 0.4279$
35	Pineal body	$-5.4813 \times 10^{-10}f + 30.7$	11.1	5.0	23.7	0.20	7.6	18.4	0.76	$8.0409 \times 10^{-5}f^{0.5} - 0.5033$
36	Placenta	$-5.2593 \times 10^{-10}f + 29.5$	19.8	6.4	25.2	0.16	6.0	17.7	1.18	$8.0909 \times 10^{-5}f^{0.5} - 0.88451$
37	Pons	$-4.6206 \times 10^{-10}f + 25.1$	55.0	5.2	18.1	0.15	2.9	17.9	0.64	$6.977 \times 10^{-5}f^{0.5} - 0.94705$
38	Prostate	$-5.2654 \times 10^{-10}f + 30.5$	20.5	5.8	23.8	0.16	6.5	18.3	0.85	$8.1579 \times 10^{-5}f^{0.5} - 0.72014$
39	Salivary gland	$-5.3099 \times 10^{-11}f + 69.8$	3.8	3.3	2.3	0.43	3.3	19.1	0.68	$9.8311 \times 10^{-6}f^{0.5} - 0.069134$
40	Seminal vesicle	$-5.1902 \times 10^{-10}f + 30.4$	18.9	5.8	23.8	0.17	6.5	18.3	0.86	$8.2466 \times 10^{-5}f^{0.5} - 0.82903$
41	Skull	$-4.6119 \times 10^{-11}f + 5.64$	3.3	2.0	4.8	0.21	3.1	11.8	0.056	$8.7997 \times 10^{-6}f^{0.5} - 0.1548$
42	Skull cancellous	$-8.5937 \times 10^{-11}f + 8.48$	7.5	4.4	7.6	0.24	3.5	12.4	0.156	$1.6762 \times 10^{-5}f^{0.5} - 0.32434$
43	Thalamus	$-4.8008 \times 10^{-11}f + 27$	37.6	5.4	20.0	0.15	3.6	18.1	0.45	$7.467 \times 10^{-5}f^{0.5} - 0.93006$
44	Tooth	$-5.0423 \times 10^{-11}f + 5.77$	3.1	2.0	4.6	0.22	3.4	11.8	0.057	$8.9656 \times 10^{-6}f^{0.5} - 0.11706$
45	Ureter\urethra	$-3.3084 \times 10^{-10}f + 21.4$	21.6	4.5	18.4	0.15	6.6	16.0	0.40	$5.6855 \times 10^{-5}f^{0.5} - 0.82104$
46	Urinary Bladder Wall	$-1.3949 \times 10^{-10}f + 9.56$	4.4	1.8	7.4	0.21	6.5	16.0	0.28	$2.1698 \times 10^{-5}f^{0.5} - 0.18294$
47	Vagina	$-4.0261 \times 10^{-10}f + 27.1$	30.8	7.9	22.2	0.17	4.9	18.1	0.60	$7.8481 \times 10^{-5}f^{0.5} - 1.5547$
48	Vertebrae	$-4.8953 \times 10^{-11}f + 5.74$	3.0	2.0	4.6	0.25	3.4	11.8	0.058	$9.4144 \times 10^{-6}f^{0.5} - 0.17538$

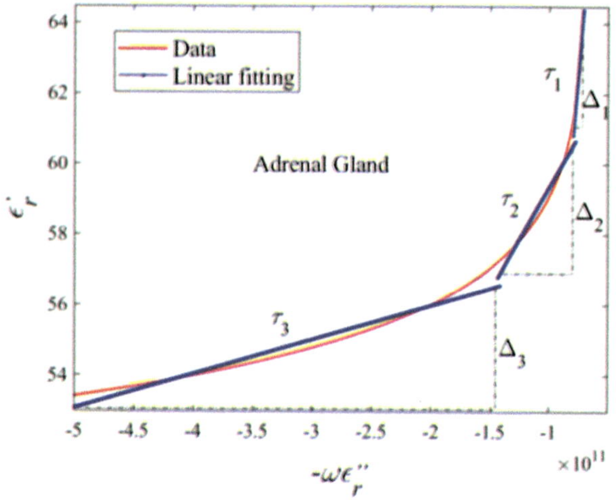

Fig. 2.7 Plot of ε_r' versus $-\omega\varepsilon_r''$ for the adrenal gland over the frequency range from 100 MHz to 20 GHz

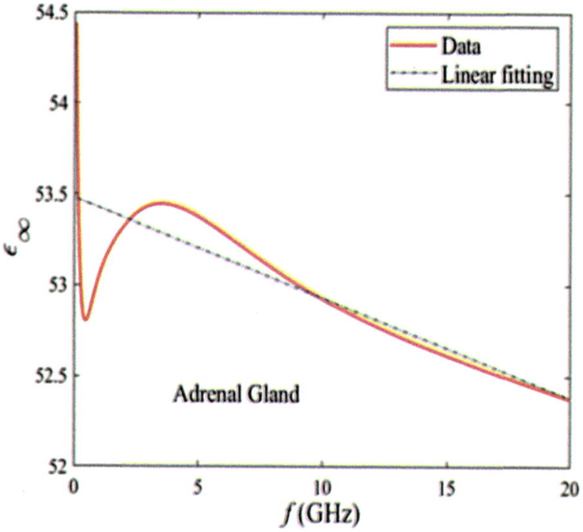

Fig. 2.8 The frequency-dependent ε_∞ of the adrenal gland

It is observed that the value of ε_∞ demonstrates an almost linear relationship with the operating frequency, f, as depicted in Fig. 2.8. The corresponding values of ε_r' in Eq. (2.5) are extracted from the literature [6]. For instance, the value of ε_∞ for adrenal gland tissue is expressed through an empirical linear equation as:

$$\varepsilon_\infty = -5.4983 \times 10^{-11} f + 53.5 \tag{2.6}$$

Step 3:
Starting from step 3, the study modeling is focused on the imaginary part of permittivity, ε_r'' (the loss factor). Despite having determined values for Δ_1, Δ_2, Δ_3, f_{r_1}, f_{r_2}, f_{r_3}, and ε_∞, a significant deviation between the ε_r'' value obtained from the Debye model and the ε_r'' data extracted from the literature [6] is observed. Consequently, DC conductivity, σ term is introduced and added in the Debye model. In step 3, the value of σ is fine-tuned to achieve a close match with the extracted ε_r'' data at a low operating frequency (DC stage). For adrenal gland tissue, a value of $\sigma = 0.63$ S/m provides a good fit for frequencies below 1 GHz, as shown in Fig. 2.9.

Step 4:
While the σ term has been incorporated into the Debye model, a minor deviation persists between the ε_r'' value derived from the Debye model and the ε_r'' data obtained from [6] at high frequencies ($f > 3$ GHz) as indicated in Fig. 2.9. Hence, to enhance the modeling accuracy in ε_r'', the error, ζ term due to dielectric loss at high frequencies is introduced. It is anticipated that high-frequency loss, ζ is proportionate to the square root of frequency, as shown in Fig. 2.10. Hence, the ζ term of the adrenal gland tissue is expressed as:

$$\zeta = 1.5458 \times 10^{-5}\sqrt{f} - 0.4735 \tag{2.7}$$

In conclusion, the comprehensive Debye model for adrenal gland tissue can be formulated as:

$$\varepsilon_r = \left\{ \begin{array}{l} \left(-5.4983 \times 10^{-11} f + 53.5\right) + \frac{3.93}{1+\left(j\frac{f}{0.68\times 10^9}\right)} + \frac{3.6}{1+\left(j\frac{f}{2.7\times 10^9}\right)} \\ + \frac{2.4}{1+\left(j\frac{f}{16.48\times 10^9}\right)} - j\frac{0.63}{2\pi\varepsilon_o f} - j\left(1.5458 \times 10^{-5}\sqrt{f} - 0.4735\right) \end{array} \right\} \tag{2.8}$$

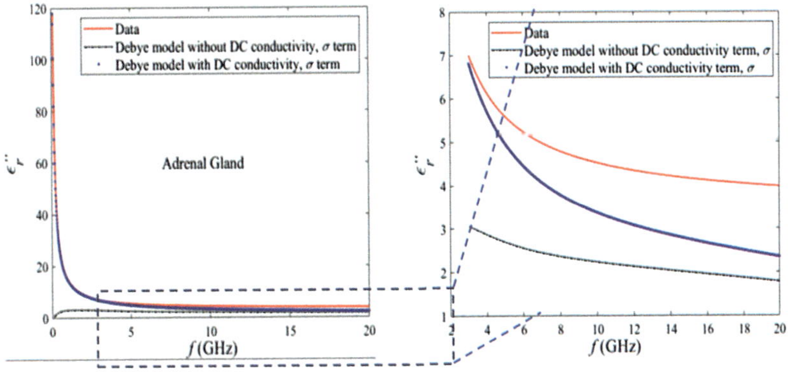

Fig. 2.9 Comparing ε_r'' values among data extracted from [6], the Debye model with and without the DC conductivity term, σ

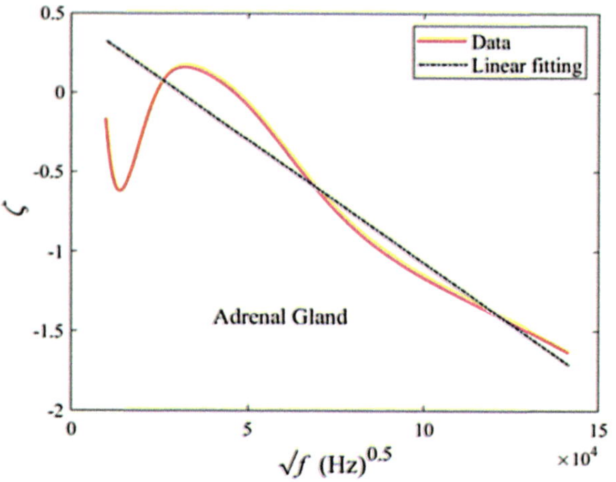

Fig. 2.10 High-frequencies dielectric loss, ζ values as functions of the square root of the operating frequency

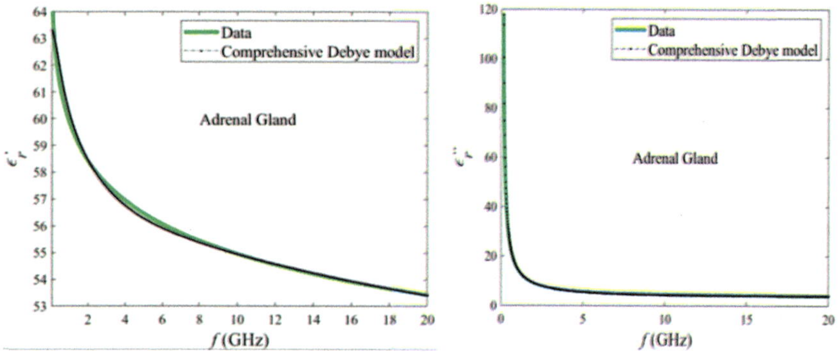

Fig. 2.11 Variation in ε_r' and ε_r'' with frequency for adrenal gland tissue

The comparison between the extracted ε_r data and the ε_r values calculated using Eq. (2.8) is presented in Fig. 2.11, demonstrating their strong agreement. The permittivity modeling process described above has been applied to a variety of tissue/organ types. The parameters for the comprehensive Debye model for all 48 types of tissues/organs are presented in Table 2.3.

2.2 Artificial Human Phantoms

The artificial human phantoms are designed to mimic the anatomical and physiological properties of the human body and simulate its interactions with RF/microwave radiation. They find applications in fields like microwave hyperthermia treatment, electromagnetic field exposure studies, medical imaging, and communication product testing.

The human phantoms can be constructed from a range of materials, such as plastics, polymers, gels, and tissue-equivalent substances, effectively mimicking the electrical, thermal, and mechanical properties of human tissues. Currently, a variety of methods have been documented [7–13] for crafting human phantoms. These phantoms can be tailored for specific human body parts, including the head, torso, limbs, and individual tissues or organs. In this chapter, the author provides a brief discussion focusing on human head and breast phantoms, as well as synthetic body fluids.

2.2.1 Human Head Phantom

A human head phantom is a specialized type of artificial human phantom designed to simulate the anatomical and physiological characteristics of the human head and brain. Human head phantom can be used:

(a) To calibrate and test medical imaging devices like MRI scanners, computed tomography (CT) scanners, and positron emission tomography (PET) machines, ensuring the accuracy and safety of diagnostic imaging.
(b) To simulate the delivery of radiation to brain tumors. This helps radiation oncologists plan treatment and assess the impact on healthy brain tissue.
(c) To study brain activity, such as functional MRI (fMRI) studies. These phantoms allow for controlled experiments to better understand brain function and neurological disorders.
(d) To test the performance and safety of mobile phones and wireless devices, ensuring compliance with safety regulations and preventing excessive radiation exposure to users.

This section introduces the formulation preparation of a human head phantom fluid (liquid contained in the head) based on IEEE Standard 1528, as shown in Table 2.4.

2.2.2 Human Breast Phantom

In the past 20 years, microwave synthetic breast tissue sample preparation methods have been largely reported by researchers. In 2011, Romeo et al. [11] presented a simple synthetic breast that mimicked the dielectric properties of the tumor, 0–30% adipose, 31–84% adipose, and 85–100% adipose of the breast using the Cole-Cole parameters in Table 2.5. The preparation of samples could be prepared by mixing different ratios of water and TX-100 as listed in Table 2.5. The recommended operating frequency range for the synthetic breast tissues is 0.5–6 GHz. Symbol T is the solution to represent the tumor breast tissues. The $B1$, $B2$, and $B3$ represent the solutions for three categories of normal breast tissues, respectively. The certain ratio volume (based on Table 2.5) of the Triton X-100 (TX-100) solution and water

Table 2.4 Human head phantom recipe based on IEEE standard 1528 [13]

f (GHz)	Ref.	Materials (% by total weight)							
		Bacteri-cide	Diacetin	DGBE	HEC	NaCl (salt)	Sucrose (sugar)	Triton X-100	Water
0.45	[8]	0.19	–	–	0.98	3.95	56.32	–	38.56
	[9]	0.5	48.9	–	–	1.7	–	–	48.9
0.835	[8]	0.1	–	–	1.0	1.45	57.0	–	40.45
0.9	[8]	0.1	–	–	1.0	1.48	56.5	–	40.92
	[9]	0.5	49.2	–	–	1.1	–	–	49.2
1.45	[8]	–	–	45.41	–	0.67	–	–	53.8
1.8	[8]	–	–	47.0	–	0.36	–	–	52.64
	[10]	–	–	13.84	–	0.35	–	30.45	55.36
	[10]	–	–	44.92	–	0.18	–	–	54.9
	[9]	0.5	49.43	–	–	0.64	–	–	49.43
1.9	[8]	–	–	44.94	–	0.18	–	–	54.9
	[10]	–	–	13.84	–	0.35	–	30.45	55.36
2.1	[8]	–	–	50.0	–	–	–	–	50.0
	[10]	–	–	7.99	–	0.16	–	19.97	71.88
2.45	[10]	–	–	7.99	–	0.16	–	19.97	71.88
	[9]	–	–	0.5	49.75	–	–	–	49.75
3	[10]	–	–	7.99	–	0.16	–	19.97	71.88

Note: *DGBE* diethylene glycol butyl ether, *HEC* hydroxyethyl cellulose, *NaCl* sodium chloride

Table 2.5 Synthetic breast tissues [19]

Symbol	TX-100 (% volume)	Water (% volume)
T	20	80
$B1$	30	70
$B2$	40	60
$B3$	100	0

Table 2.6 Synthetic human blood plasma (for 1.0 L) [12]

No.	Reagents	Amount (g)
1	Sodium chloride, NaCl (99.5%)	8.036
2	Sodium hydrogen carbonate, $NaHCO_3$ (99.5%)	0.352
3	Potassium chloride, KCl (99.0%)	0.225
4	Sodium dihydrogen phosphate dihydrate, $Na_2HPO_4.2H_2O$ (99.5%)	0.230
5	Magnesium chloride hexahydrate, $MgCl_2{\cdot}6H_2O$ (98%)	0.311
6	Calcium chloride dihydrate, $CaCl_2.2H_2O$ (99%)	0.293
7	Sodium sulphate, Na_2SO_4	0.072
8	Tris (hydroxymethyl) aminometane, $(CH_2OH)_3CNH_2$ (99.2%)	6.063
9	Hydrochloric fuming 37%, HCl (37 vol%)	<30 mL

are measured using a volumetric burette. Later, both of the solutions are poured and mixed into a conical flask. The samples of T and $B1$ were prepared at room temperature, while the solution of $B2$ was prepared at 60 °C due to the high viscosity of the mixture solution [19].

2.2.3 *Human Blood Plasma (Synthetic Body Fluid)*

Synthetic body fluid, also known as artificial human blood plasma, is a laboratory-created solution that replicates the chemical composition and characteristics of natural human blood plasma. Synthetic body fluid is formulated to replicate the major components of natural human blood plasma, including water, electrolytes (such as sodium, potassium, calcium, and chloride), proteins (such as albumin and globulins), hormones, gases (such as oxygen and carbon dioxide), and waste products (such as urea and creatinine). This study introduces one of the preparation recipes for synthetic body fluid, which involves a mixture of chemical reagents.

The powder form reagents, their purity, and the quantities used in the preparation of 1.0-L synthetic human blood plasma (SHBP) are listed in Table 2.6. The aqueous Hydrochloric (HCl) is used to adjust the synthetic plasma to a pH of 7.40 at 36.5 °C. First, about 700 mL of ultra-pure water is poured into a 1000 mL polypropylene beaker, all powder form reagents (according to their respective amounts) are mixed with water in the beaker and the fluid mixture was stirred using a magnetic bar at 36.5 °C. After each preceding reagent had completely dissolved, the fluid then is transferred to a 1000 mL volumetric Pyrex_ glass flask, cooled to approximately

20 °C, and the total volume adjusted to 1.0 L by adding ultra-pure water. Besides, the pH of the fluid is adjusted to 7.4 by adding a little HCl.

2.3 Assessing RF/Microwave Exposure Safety in WBANs

2.3.1 Penetration Depth

Penetration depth, D essential for WBAN design, determines signal attenuation rates through tissue. A shorter penetration depth, D implies faster attenuation, crucial for reliable in-body communication. In addition, the penetration depth, D also aids in assessing effective communication ranges, and optimizing network coverage. In healthcare, it assists diagnostics by revealing how signals interact with specific tissues and organs.

The penetration depth, D (in unit meters) of tissues/organs exposed to microwave radiation can be expressed in terms of ε_r' and ε_r'' as:

$$D \approx \frac{\lambda_o \sqrt{\varepsilon_r'}}{2\pi \varepsilon_r''} \tag{2.9}$$

where λ_o is the wavelength of the supplied microwave. The penetration depth, D can vary among different types of tissues/organs due to differences in their dielectric properties. Typically, industrial, scientific, and medical (ISM) applications utilize the operating frequencies of 915 MHz and 2.45 GHz, which are allocated by the US Federal Communications Commission (FCC). In recent research, some experts have explored the use of a 5.8 GHz operating frequency for ISM applications [20].

2.3.2 Specific Absorption Rate (SAR) Analysis

Specific absorption rate (SAR) is a measure of the rate at which energy is absorbed by unit mass of biological tissues when exposed to RF and microwave radiation [15]. It is typically expressed in watts per kilogram (W/kg). The SAR of the biological tissues/organs can be defined as:

$$\text{SAR} = \frac{\sigma}{2\rho}|E|^2 = \frac{2\pi f \varepsilon_o \varepsilon_o''}{2\rho}|E|^2 \tag{2.10}$$

where σ is the conductivity of the human tissues (S/m). The ρ is the mass density of the human tissues (kg/m^3). The E is the time-varying peak intensity of the electric field within the tissues (V/m). In addition, The SAR ties in closely with the tissues/organs temperature, T (in °C), which can be written as:

Table 2.7 Simulated SAR for on-body rectangular patch antennas with a gain of 2.6 dBi at 2.4 GHz [21]

Case	Sensor node position	1 g tissue (W/kg)	10 g tissue (W/kg)
1	Left wrist	0.214	0.105
2	Left chest	0.0514	0.0188
3	Left index	0.088	0.0495
4	Left tight	0.0795	0.041
5	Head	0.47	0.201

$$\text{SAR} = C_\rho \frac{\partial T}{\partial t} \approx C_\rho \frac{\Delta T}{\Delta t} \tag{2.11}$$

where C_ρ is the specific heat capacity of the tissues/organs in unit of J/(kg °C), and t is the time (s). On the other hand, T is the temperature, which varies spatially and temporally within the tissue. From Eqs. (2.10) and (2.11), the value of absorbed power, P by the tissues can be estimated as:

$$P \approx C_\rho \times \rho \times \frac{\Delta T}{\Delta t} \tag{2.12}$$

Once $\Delta T/\Delta t$, ρ, and ε_r'' are known, the absorbed power, P can be determined from Eq. (2.12).

The SAR measurements are influenced by the volume or mass of the tissue being tested. Typically, fixed mass values like 1 g or 10 g within cubic shapes are used. There are two primary radiation protection standards: one established by the European Committee for Electrotechnical Standardization (CENELEC), joined by at least 33 European countries, and the other by the Federal Communications Commission (FCC). The CENELEC standard sets a maximum SAR limit for commercially sold devices at 2 W/kg averaging over 10 g of body tissue, while the FCC standard limits it to 1.6 W/kg averaging over 1 g of body tissue. Typically, thermal biological effects from RF and microwave radiation occur when SAR values exceed the 4 W/kg limit, leading to a tissue heating of 1 °C (general population limit: SAR 0.08 W/kg).

Marwen et al. [21] investigated SAR using a voxel-based human body model (Katja model). The coordinator (receiver) is positioned at the front center of the body's waist. Five sensor nodes (transmitters) placed at different parts of the body are simulated as depicted in Fig. 2.12. Both the coordinator and sensor nodes utilized identical rectangular microstrip patch antennas with a gain of 2.6 dBi at 2.4 GHz. The ground part of the antenna was attached to the body's surface, and the front radiation part of the antenna extended forward from the body's surface. The simulated SAR for the sensor nodes at five different positions is respectively shown in Table 2.7.

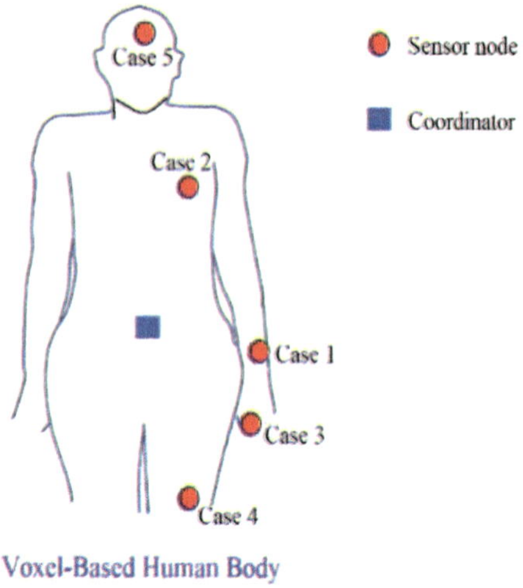

Fig. 2.12 Communication links between two rectangular patch antennas (coordinator and sensor node) with a gain of 2.6 dBi at 2.4 GHz [21]

2.3.3 *Bio-heat Equations*

Bio-heat equations are the governing equations used to simulate how temperature distributes within biological tissues when they are exposed to external heat sources, such as RF energy emitted by WBAN devices. Several bio-heat equations are available, but the Pennes bioheat transfer equation is the most commonly used. It is expressed as:

$$\rho c \frac{\partial T}{\partial t} = \nabla \cdot (k \nabla T) + \rho_b c_b \omega_b (T_a - T) + Q_{met} + Q \tag{2.13}$$

where ρ_b, c_b, and ω_b are the blood density (in kg/m^3), the specific heat capacity of blood (in J/kg/°C), and the blood perfusion rate (in mL/min/kg), respectively. On the other hand, the ρ, c, and k are the mass density of the tissues/organs (in kg/m^3), the specific heat of the tissue (in J/kg/°C), and the thermal conductivity (in W/m/°C), which indicates how heat is conducted within the tissue. While T_a (= 36.5 °C) is the supplying arterial blood temperature (in °C).

The $\nabla \cdot (k\nabla T)$ term in Eq. (2.13) represents heat conduction within the tissue. It accounts for the flow of heat from regions of higher temperature to regions of lower temperature. The thermal conductivity, k plays a significant role in determining how effectively heat is conducted through the tissue. For instance, if the heat flow propagates forward symmetrically in the radial, r and frontal, z directions, the $\nabla \cdot (k\nabla T)$ term can be written as [22]:

$$\nabla \cdot (k\nabla T) = k\left(\frac{\partial^2 T}{\partial r^2} + \frac{1}{r}\frac{\partial T}{\partial r} + \frac{\partial^2 T}{\partial z^2}\right) \tag{2.14}$$

On the other hand, the Q_{met} represents the heat generated within the tissue due to metabolic processes, such as cellular respiration. This Q_{met} term considers the heat produced by the body's natural physiological functions. In addition, Q represents heat added to the tissue or heat absorbed by tissue from external sources, such as IoBTs, medical devices or therapies. This is particularly important in medical applications where the goal may be to raise the temperature, T in specific tissue regions. The relationship between Q value in (2.13) and SAR from Eq. (2.10) is given as:

$$\text{SAR} = \frac{2\pi f \varepsilon_o \varepsilon_o''}{2\rho}|E|^2 = \frac{Q}{\rho} \tag{2.15}$$

Hence, once the SAR was known, the Q can be determined as:

$$Q = \rho \times \text{SAR} \tag{2.16}$$

In general, numerical methods, such as finite-difference (FD), finite-difference time-domain (FDTD), finite element (FE), or finite volume (FV) methods, are used to discretize the Eq. (2.13) (partial differential equation nature) and solve it over time and space. These methods break down the partial differential equation (PDE) into smaller steps, allowing for iterative solutions [22]. For instance, using FD method, the (2.14) can be discretized as:

$$k\left(\frac{\partial^2 T}{\partial r^2} + \frac{1}{r}\frac{\partial T}{\partial r} + \frac{\partial^2 T}{\partial z^2}\right) = k\begin{bmatrix} \frac{T(i+1,j)-2T(i,j)+T(i-1,j)}{\Delta r^2} \\ +\frac{1}{r}\frac{T(i+1,j)-T(i,j)}{2\Delta r} \\ +\frac{T(i,j+1)-2T(i,j)+T(i,j-1)}{\Delta z^2} \end{bmatrix} \tag{2.17}$$

where Δr and Δz are the discretized grid size in the radial, r and frontal, z directions, respectively. The temperature, T evaluated at the specified point (r, z) is defined as $T(i, j)$. For a point where distance Δr is shifted to the right, the temperature at that point is written as $T(i, j + 1)$. Similarly, for a point where distance Δr is shifted to the left, the temperature is $T(i, j - 1)$. The parameters specified in (2.13) for different tissues/organs are detailed in Table 2.8.

Table 2.8 Specific absorption rate (SAR) and bioheat transfer parameters for various human body tissues/organs [6]

No.	Body tissues/organs	Density (kg/m^3)	Heat capacity (J/kg/°C)	Thermal conductivity (W/m/°C)	Heat transfer rate (mL/min/kg)	Heat generation rate (W/kg)
1	Adrenal gland	1028	3513	0.44	1458	22.58
2	Bile	928	4037	0.58	0	0
3	Blood	1050	3617	0.52	10,000	0
4	Blood plasma	1020	3930	0.58	–	NaN
5	Blood serum	1024	–	–	–	NaN
6	Blood vessel wall	1102	3306	0.46	150	2.32
7	Bone (cancellous)	1178	2274	0.31	30	0.46
8	Bone (cortical)	1908	1313	0.32	10	0.15
9	Bone marrow (red)	1029	2666	0.28	135	2.09
10	Bone marrow (yellow)	980	2065	0.19	30	0.46
11	Brain	1046	3630	0.51	559	11.37
12	Brain (grey matter)	1045	3696	0.55	764	15.54
13	Brain (white matter)	1041	3583	0.48	212	4.32
14	Breast fat	911	2348	0.21	47	0.73
15	Breast gland	1041	2960	0.33	150	2.32
16	Bronchi	1102	3306	0.46	238	3.69
17	Bronchi lumen	1	1004	0.03	0	0
18	Cartilage	1100	3568	0.49	35	0.54
19	Cerebellum	1045	3653	0.51	770	15.67
20	Cerebrospinal fluid	1007	4096	0.57	0	0

21	Cervix	1105	3676	0.53	700	10.84
22	Commissura anterior	1041	3583	0.48	212	4.32
23	Commissura posterior	1041	3583	0.48	212	4.32
24	Connective tissue	1027	2372	0.39	37	0.58
25	Diaphragm	1090	3421	0.49	99	2.44
26	Ductus deferens	1102	3306	0.46	188	2.91
27	Dura	1174	3364	0.44	380	5.89
28	Epididymis	1082	3778	0.52	200	3.09
29	Esophagus	1040	3500	0.53	190	2.94
30	Esophagus lumen	1	1004	0.03	0	0
31	Extracellular fluids	1006	3997	0.60	0	0
32	Eye (aqueous humor)	994	3997	0.58	0	0
33	Eye (choroid)	1050	3617	0.52	5950	92.15
34	Eye (ciliary body)	1090	3421	0.49	3800	93.72
35	Eye (cornea)	1062	3615	0.54	0	0
36	Eye (iris)	1090	3421	0.49	1632	40.26
37	Eye (lens)	1076	3133	0.43	0	0
38	Eye (retina)	1036	3696	0.55	240	4.89
39	Eye (sclera)	1032	4200	0.58	380	5.89
40	Eye (vitreous humor)	1005	4047	0.59	0	0
41	Eye lens (cortex)	1076	3133	0.43	0	0
42	Eye lens (nucleus)	1076	3133	0.43	0	0
43	Fat	911	2348	0.21	33	0.51
44	Fat (average infiltrated)	911	2348	0.21	33	0.51

(continued)

Table 2.8 (continued)

No.	Body tissues/organs	Density (kg/m^3)	Heat capacity (J/kg/°C)	Thermal conductivity (W/m/°C)	Heat transfer rate (mL/min/kg)	Heat generation rate (W/kg)
45	Fat (not infiltrated)	911	2348	0.21	0	0
46	Gallbladder	1071	3716	0.52	30	0.46
47	Heart lumen	1050	3617	0.52	10,000	0
48	Heart muscle	1081	3686	0.56	1026	39.45
49	Hippocampus	1045	3696	0.55	764	15.54
50	Hypophysis	1053	3687	0.51	885	13.71
51	Hypothalamus	1045	3696	0.55	764	15.54
52	Intervertebral disc	1100	3568	0.49	35	0.54
53	Kidney	1066	3763	0.53	3795	18.05
54	Kidney (cortex)	1049	3587	0.53	3874	18.43
55	Kidney (medulla)	1044	3745	0.54	599	2.85
56	Large intestine	1088	3655	0.54	765	11.85
57	Large intestine lumen	1045	3801	0.56	0	0
58	Larynx	1100	3568	0.49	35	0.54
59	Liver	1079	3540	0.52	860	9.93
60	Lung	394	3886	0.39	401	6.21
61	Lung (deflated)	1050	3886	0.39	401	6.21
62	Lung (inflated)	394	3886	0.39	401	6.21
63	Lymph	1019	–	–	0	0
64	Lymphnode	1035	3739	0.46	461	7.15
65	Mandible	1908	1313	0.32	10	0.15
66	Medulla oblongata	1046	3630	0.51	559	11.37

67	Meniscus	1100	3568	0.49	35	0
68	Midbrain	1046	3630	0.51	559	11.37
69	Mucous membrane	1102	3150	0.34	594	9.19
70	Muscle	1090	3421	0.49	37	0.91
71	Nerve	1075	3613	0.49	160	2.48
72	Ovary	1048	3778	0.52	236	3.65
73	Pancreas	1087	3164	0.51	767	11.89
74	Penis	1102	3306	0.46	12	0.19
75	Pharynx	1	1004	0.03	0	0
76	Pineal body	1053	3687	0.51	885	13.71
77	Placenta	1018	3807	0.52	1700	26.33
78	Pons	1046	3630	0.51	559	11.37
79	Prostate	1045	3760	0.51	394	6.10
80	SAT (subcutaneous fat)	911	2348	0.21	33	0.51
81	Salivary gland	1048	3760	0.51	383	5.93
82	Scalp	1109	3391	0.37	106	1.65
83	Seminal vesicle	1045	3760	0.51	394	6.10
84	Skin	1109	3391	0.37	106	1.65
85	Skull	1908	1313	0.32	10	0.15
86	Skull cancellous	1178	2274	0.31	30	0.46
87	Skull cortical	1908	1313	0.32	10	0.15
88	Small intestine	1030	3595	0.49	1026	15.89
89	Small intestine lumen	1045	3801	0.56	0	0
90	Spinal cord	1075	3630	0.51	160	2.48

(continued)

Table 2.8 (continued)

No.	Body tissues/organs	Density (kg/m^3)	Heat capacity (J/kg/°C)	Thermal conductivity (W/m/°C)	Heat transfer rate (mL/min/kg)	Heat generation rate (W/kg)
91	Spleen	1089	3596	0.53	1557	24.11
92	Stomach	1088	3690	0.53	460	7.13
93	Stomach lumen	1045	3801	0.56	0	0
94	Tendon\ligament	1142	3432	0.47	29	0.45
95	Testis	1082	3778	0.52	200	3.09
96	Thalamus	1045	3696	0.55	685	13.93
97	Thymus	1023	3043	0.34	247	3.83
98	Thyroid gland	1050	3609	0.52	5624	87.10
99	Tongue	1090	3421	0.49	78	1.21
100	Tooth	2180	1255	0.59	0	0
101	Tooth (dentine)	2063	1338	0.52	0	0
102	Tooth (enamel)	2958	711	0.79	0	0
103	Trachea	1080	3568	0.49	35	0.54
104	Trachea lumen	1	1004	0.03	0	0
105	Ureter\urethra	1102	3306	0.46	188	2.91
106	Urinary bladder wall	1086	3581	0.52	78	1.21
107	Urine	1024	4178	0.56	0	0
108	Uterus	1105	3676	0.53	458	7.09
109	Vagina	1088	3655	0.54	98	1.52
110	Vertebrae	1908	1313	0.32	10	0.15
111	Water	994	4178	0.60	0	0

2.4 Wireless Human Body Area Networks (WBANs)

Wireless human body area networks (WBANs) enable real-time monitoring and data transmission for various medical, healthcare, and wellness applications. Nevertheless, WBAN technology encounters various challenges, including data privacy and security, interoperability, and IoBT battery life. Hence, efficiently safeguarding sensitive health data, ensuring seamless communication among diverse WBAN devices, and improving IoBT battery life are currently vital concerns in modern healthcare.

At present, IEEE 802.15.6-2012 is the established standard governing wireless medical devices within Wireless Body Area Networks (WBANs) [13]. Three main distinct frequency bands are allocated for establishing four primary communication links among IoBTs within the WBAN, as illustrated in Fig. 2.13. The frequency range of 402–405 MHz is designated for narrowband medical implant communication services, specifically for in-body (implanted) to on-body (surface) communication. Conversely, the 2.4–2.4835 GHz frequency range (within the ISM band) is employed for on-body to on-body and on-body to external communications. Besides, the IEEE 802.15.6-2012 standard defines the 3.1–10.6 GHz ultra-wideband (UWB) spectrum to enable on-body to on-body and on-body to external communications, facilitating low power consumption, compact antennas, and high data rates.

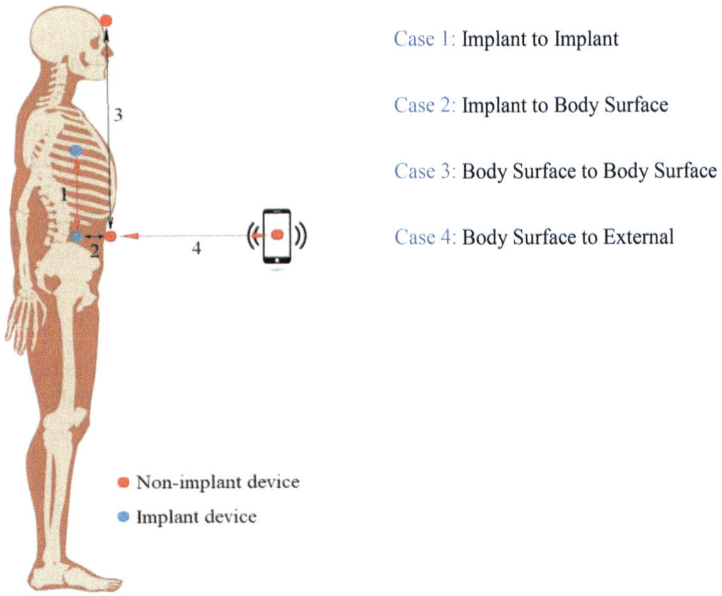

Fig. 2.13 Possible case channels for WBAN

2.4.1 In-Body Channel Modeling

In-body channel modeling provides the theoretical and practical foundation for the development and optimization of WBANs [23–26]. In-body channel modeling ensures reliable wireless communication for implantable medical devices like pacemakers, neurostimulators, and drug delivery systems. The modeling considers signal propagation factors like absorption (attenuation), spreading (dispersion), and reflections within the human body. In addition, different tissues within the human body have varying dielectric properties that affect signal propagation, hence statistical techniques are often used to analyze this variability. The in-body path loss, *PL* in dB can be generally expressed as:

$$PL\,(\text{dB}) = PL_{Absorption}\,(\text{dB}) + PL_{Spreading}\,(\text{dB}) \tag{2.18}$$

where $PL_{Absorption}$ and $PL_{Spreading}$ are the path losses due to absorption and spreading losses, respectively. A practical in-body path loss model in unit dB can be empirically described as [23]:

$$PL\,(\text{dB}) = PL_o\,(\text{dB}) + 10n\log_{10}\left(\frac{d}{d_o}\right) + S \tag{2.19}$$

where d_o is the reference distance and PL_o is the mean path loss at a reference distance d_o. Symbol *n* is the path loss exponent which depends on the internal body environment where RF signal is propagating. On the other hand, S [$\sim N(\mu, \sigma_s^2)$] is the random variable, so-called shadowing, which has a normal distribution with zero mean ($\mu = 0$) and a certain standard deviation, σ_s^2. In fact, *S* is the deviation term (in dB) due to different kinds of tissues/organs in the human body, movement of the body parts, and wireless signal radiated in different directions [line of sight (LOS) and non-line of sight (NLOS)].

Furthermore, the in-body path loss, *PL* model in unit dB, can be simplified as [24]:

$$PL\,(\text{dB}) = a\log_{10}(d) + b + S \tag{2.20}$$

where *a* and *b* are the linear fitting coefficients. On the other hand, *d* is the distance between the transmitter and receiver in unit mm. Symbol *S* is the normal distribution variable with zero mean ($\mu = 0$) and standard deviation, σ_s^2 (Tables 2.9, 2.10, 2.11, and 2.12; Figs. 2.14 and 2.15).

Table 2.9 Path loss, *PL* (dB) parameters in (2.19) for 402–405 MHz

Communication link	Location	PL_o (dB)	d_o (mm)	n	μ (dB)	σ_s^2 (dB)
Implant to implant	Deep tissue	35.04	50	6.26	0	8.18
	Near surface	40.94	50	4.99	0	9.05
Implant to body surface	Deep tissue	47.14	50	4.26	0	7.85
	Near surface	49.81	50	4.22	0	6.81

Table 2.10 Path loss, *PL* (dB) parameters in (2.20) for body surface to body surface case [24]

Communication link	Frequency range, f (MHz)	Test environment	a	b	μ (dB)	σ_s^2 (dB)
Body surface to body surface	400–450	Hospital room	3	34.6	0	4.63
		Anechoic chamber	22.6	−7.85	0	5.60
	608–614	Hospital room	16.7	−0.45	0	5.99
		Anechoic chamber	17.2	1.61	0	6.96
	950–956	Hospital room	15.5	5.38	0	5.35
		Anechoic chamber	28.8	−23.5	0	11.7
	2400–2500 (ISM)	Hospital room	6.6	36.1	0	3.80
		Anechoic chamber	29.3	−16.8	0	6.89
	3100–10,600 (UWB)	Hospital room	19.2	3.38	0	4.40
		Anechoic chamber	34.1	−31.4	0	4.85

Table 2.11 Path loss, *PL* (dB) parameters in (2.19) for 3.1–10.6 GHz

Communication link	Position/direction	Antenna separation from body surface, d (mm)	PL_o (dB)	d_o (mm)	n	μ (dB)	σ_s^2 (dB)
Body surface to body surface	Around torso	0	56.1	100	5.8	0	0
		5	48.4	100	5.9	0	0
		10	45.8	100	6.0	0	0
	Along torso	0	56.5	100	3.1	0	0
		5	44.6	100	3.1	0	0

Table 2.12 Path loss, *PL* (Eq. 2.19) for the implant to surface body from 3.1 to 5.1 GHz, which only valid for a maximum distance, $d = 79.5$ mm [25]

Case	PL_o (dB)	d_o (mm)	n	μ (dB)	σ_s^2 (dB)	Root mean squared estimator (RMSE)
In vivo measurement	26.2266	10.0	5.3967	0	4.4972	28.1
CST simulation (Nelly model)	−8.3767	10.0	8.9681	0	1.9795	7.5
Measurement using phantom	−29.7593	10.0	10.3395	0	2.3069	3.64
Measurement using phantom (shifted)	6.29	10.0	7.3824	0	2.3772	3.88

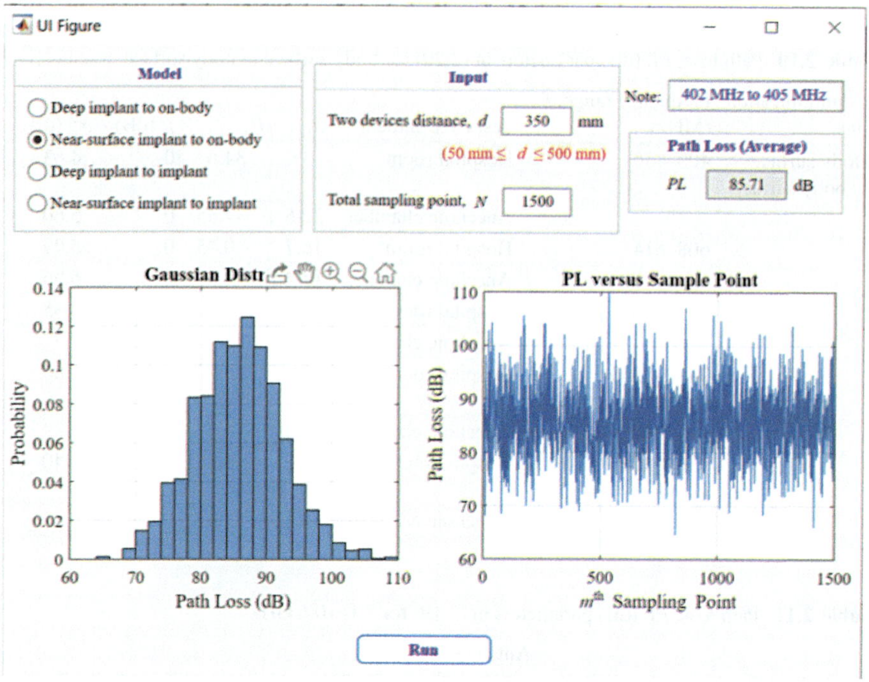

Fig. 2.14 MATLAB-based user interface (IU) of the in-body path loss, *PL* (dB) from 402 to 405 MHz

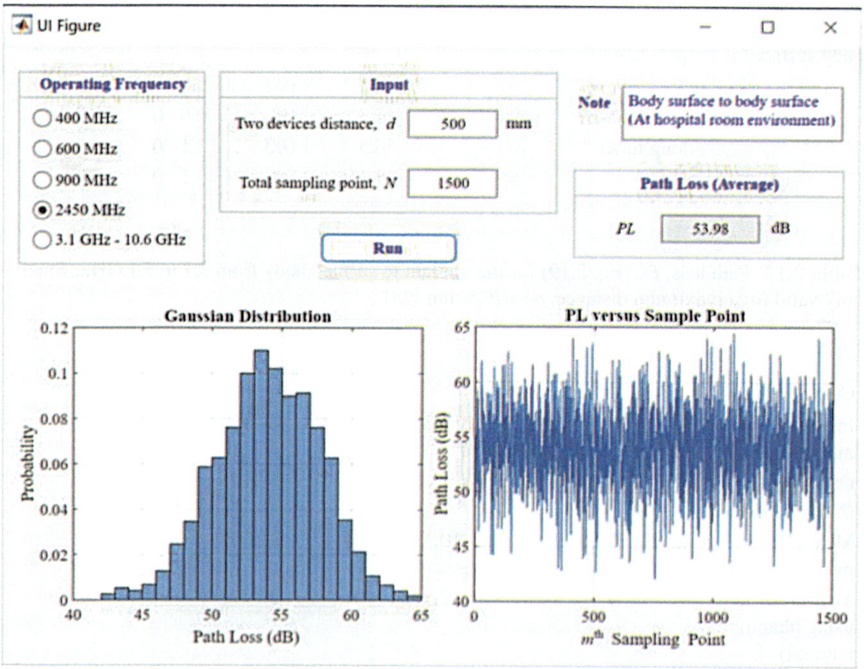

Fig. 2.15 MATLAB-based user interface (IU) of the in-body path loss, *PL* (dB) for various range of frequency spectrum

2.5 Conclusion

This chapter delves into dielectric and path loss modeling to enhance digital twins for Wireless Human Body Area Networks (WBANs). It emphasizes the pivotal role of these models in refining digital twin simulations, with applications spanning healthcare, wearable technology, and personalized medical diagnostics. As technology evolves, so does the potential of digital twins for WBANs. The methods outlined here provide a foundation for future research, with the potential to revolutionize healthcare. Moving forward, interdisciplinary collaboration and model refinement are key. The authors aim for this chapter to inspire further exploration of digital twins in WBANs, ultimately elevating healthcare and overall quality of life. These in-body channel simulations help researchers and engineers assess the performance of communication systems and design strategies to overcome channel-related challenges.

Acknowledgments This work is supported by Universiti Teknologi Malaysia (UTM) under the UTM Fundamental Research Grant(Q.J130000.3851.22H13).

References

1. A. Celik, K.N. Salama, A.M. Eltawil, The internet of bodies: A systematic survey on propagation characterization and channel modeling. IEEE Internet Things J. **9**, 321–345 (2022). https://doi.org/10.1109/JIOT.2021.3098028
2. N. Saeed, M.H. Loukil, H. Sarieddeen, T.Y. Al-Naffouri, M.-S. Alouini, Body-centric terahertz networks: Prospects and challenges. IEEE Trans. Mol. Biol. Multiscale Commun. **8**, 138–157 (2022). https://doi.org/10.1109/TMBMC.2021.3135198
3. K. Sasaki, K. Wake, S. Watanabe, Development of best fit Cole-Cole parameters for measurement data from biological tissues and organs between 1 MHz and 20 GHz. Radio Sci. **49**, 459–472 (2015). https://doi.org/10.1002/2013RS005345
4. L. Mariya, P. Dijana, M. Leah, B.W. Cynthia, J.L. Mary, H. Josephine, S. Sarah, O. Travis, M. Anthony, M.B. Tara, T. Walley, M. Daphne, H.B. John, O. Michal, C.H. Susan, A large-scale study of the ultrawideband microwave dielectric properties of normal, benign and malignant breast tissues obtained from cancer surgeries. Phys. Med. Biol. **52**, 6093–6115 (2007). https://doi.org/10.1088/0031-9155/52/20/002
5. Federal Communications Commission (FCC): Body tissue dielectric parameters tool. https://www.fcc.gov/general/body-tissue-dielectric-parameters
6. ITIS Foundation: Tissue properties. https://itis.swiss/virtual-population/tissue-properties/database/dielectric-properties/
7. S. Wood, T. Martins, T.S. Ibrahim, How to design and construct a 3D-printed human head phantom. J. 3D Print. Med. **3**, 119–125 (2019). https://doi.org/10.2217/3dp-2019-0016
8. M.Y. Kanda, M. Ballen, C.K. Chou, Formulation and characterization of tissue simulating liquids used for SAR measurement (500–2000 MHz), in *Asia-Pacific Radio Science Conference*, (Tokyo, Japan, 2001), p. 274
9. A. Peyman, C. Gabriel, Tissue equivalent liquids for SAR measurement at microwave frequencies. Poster No. P-53, Bioelectromagnetics Society Annual Meeting, Quebec, Canada (June 2002)

10. V. Vigneras, Elaboration and characterization of biological tissues equivalent liquids in the frequency range 0.9–3 GHz. Final report, Physique des Interactions Ondes-Matière (PIOM) Laboratory, University of Bordeaux, France (Nov. 2001)
11. S. Romeo, L.D. Donato, O.M. Bucci, I. Catapano, L. Crocco, M.R. Scarfì, R. Massa, Dielectric characterization study of liquid-based materials for mimicking breast tissues. Microw. Opt. Technol. Lett. **53**, 1276–1280 (2011). https://doi.org/10.1002/mop.26001
12. S. Jalota, S.B. Bhaduri, A.C. Tas, Using a synthetic body fluid (SBF) solution of 27 mM HCO_3^- to make bone substitutes more osteointegrative. Mater. Sci. Eng. C. **28**, 129–140 (2008). https://doi.org/10.1016/j.msec.2007.10.058
13. IEEE, IEEE Standard for local and metropolitan area networks–part 15.6: Wireless body area networks, in *IEEE Standard 802.15.6TM*, (IEEE, 2012)
14. IEEE, *IEEE P802.15 Working Group for Wireless Personal Area Networks (WPANs)* (IEEE P802.15-08-0780-09-0006, 2009)
15. IEEE, IEEE Recommended practice for determining the peak spatial-average specific absorption rate (SAR) in the human head from wireless communication devices: Measurement technique, in *IEEE Standard 1528-2013*, (IEEE, 2013)
16. MATLAB R2020a, MATLAB App Building. U.S: MathWorks (2020)
17. CST MPHYSICS® STUDIO: Biological data. http://www.mweda.com/cst/cst2013/mergedprojects/CST_MPHYSICS_STUDIO/common_tools/bio_models.htm#Voxel_Family
18. K.Y. You, Z. Abbas, M.F. Abdul Malek, E.M. Cheng, H.K. Mun, Modeling of dielectric relaxation for lossy materials at microwave frequencies using polynomial approaches. J. Teknol. (Sci. Eng.). **58**, 51–65 (2012). https://doi.org/10.11113/jt.v58.2544
19. C.Y. Lee, K.Y. You, Z. Abbas, K.Y. Lee, Y.S. Lee, E.M. Cheng, S-band five-port ring reflectometer-probe system for in vitro breast tumor detection. Int. J. RF Microwave Comput. Aided Eng. **28**, 1–13 (2018). https://doi.org/10.1002/mmce.21198
20. K.Y. You, Introductory chapter: RF/microwave applications, in *Emerging Microwave Technologies in Industrial, Agricultural, Medical and Food Processing*, ed. by K.Y. You, (IntechOpen, London, 2018). https://doi.org/10.5772/intechopen.73574
21. A. Marwen, H. Abdelaziz, M. Aref, Design of wearable antenna sensors and path loss modeling for internet of wearable things applications. Biomed. J. Sci. Tech. Res. **51**, 43048–43068 (2023). https://doi.org/10.26717/BJSTR.2023.51.008155
22. K.Y. You, *RF Coaxial Slot Radiators: Modeling, Measurements, and Applications* (Artech House, Norwood, 2015)
23. K. Sayrafian-Pour, W.-B. Yang, J. Hagedorn, J. Terrill, K.Y. Yazdandoost, A statistical path loss model for medical implant communication channels, in *2009 IEEE 20th International Symposium on Personal, Indoor and Mobile Radio Communications*, (IEEE Press, Tokyo, 2009), p. 1. https://doi.org/10.1109/PIMRC.2009.5449869
24. T. Aoyagi, J.-i. Takada, K. Takizawa, N. Katayama, T. Kobayashi, K.Y. Yazdandoost, H.-b. Li, R. Kohno, *Channel Model for Wearable and Implantable WBANs* (IEEE 802.15-08-0416-04-0006, November 2008)
25. P.S. Sofia, A. Carlos, G.P. Concepci6n, F. Matteo, C. Narcis, UWB path loss models for ingestible devices. IEEE Trans. Antennas Propag. **67**, 5025–5034 (2019). https://doi.org/10.1109/TAP.2019.2891717
26. S. Myrtill, M. Mats-Olof, 5G wireless communication and health effects – A pragmatic review based on available studies regarding 6 to 100 GHz. Int. J. Environ. Res. Public Health **16**, 3406 (2019). https://doi.org/10.3390/ijerph16183406

Chapter 3
ML for Digital Twin Over Wireless Networks: Creation, Deployment, and Applications

Yuchen Liu and Zhaohui Yang

3.1 Introduction

Digital twin (DT) technology is envisioned to act as an important role for modern communication society, in particular for the future Metaverse-based networks. DT is the process of using information technology to digitally define and model physical entities. The core concept is to realize feedback optimization of physical entities through the simulation, control and prediction of DTs. Although DT has been successfully applied in the fields of manufacturing [1–3] or complex system operation and maintenance [4–6], research related to the creation, deployment, and application of a digital twin for next-generation (nextG) 6G networks is still in its infancy as representing a physical network using a virtual model faces many *challenges*: (1) how to build a high-fidelity twin involving various physical objects, communication entities, signal propagation phenomenon, and network attributes synergistically; (2) how to adaptively evolve the built DTs taking into account the highly dynamic nature of nextG networks; (3) how to integrate the DT with advanced ML models to support various network tasks and applications.

By providing a living and interactive copy of a physical network system, digital twin networks (DTNs) become a promising solution and bring numerous *benefits*, such as accelerated network development processes, enhanced resource utilization, easy management, and faster networking innovation with reduced costs. A digital twin serves as a virtual representation of a physical object, system, or process.

Y. Liu (✉)
North Carolina State University, Raleigh, NC, USA
e-mail: yuchen.liu@ncsu.edu

Z. Yang
Zhejiang University, Building of College of Information Science and Electronic Engineering, Zhejiang, China
e-mail: yang_zhaohui@zju.edu.cn

M. K. Afzal et al. (eds.), *Digital Twins for Wireless Networks*,
https://doi.org/10.1007/978-3-031-73679-7_3

In the context of a digital twin network, these virtual replicas are connected to create a comprehensive network system that mirrors the interactions and behaviors of the corresponding physical entities. The composition of a digital twin network may involve multiple digital twins linked together, often in real-time, to simulate, monitor, and analyze the performance and behavior of the entire interconnected system. The goal is to create a holistic and dynamic representation that facilitates data exchange, analysis, and decision-making for improved efficiency, optimization, and understanding of the physical network system. To be specific, in the network deployment phase, the availability of DTN can deliver a nextG network planning and operation tool to see the entire life cycle of network and management details before physically placing wireless sites. If system enhancement is desired, performance parameters can be adjusted and simulated in a DTN without imperiling the operation of the entire infrastructure. Then, in the network operation phase, DTN can enhance the access and analysis of already deployed network systems to enable remote commissioning and diagnostics, which lowers service costs by remotely configuring faulty parts of a network facility, e.g., ordered and replaced accordingly for the network operator or other involved researchers; Lastly, in the network design phase, a DTN can also be used to predict network performance and behavior in complex environments by leveraging advanced ML techniques, so as to optimize facilities' efficiency and output.

There are several examples or user studies in real world scenarios when integrating ML techniques into DTNs, enhancing functionality and efficiency across various industries and complex systems. In smart manufacturing, ML-driven DTNs enable real-time monitoring and optimization of production processes. Using ML algorithms, these DTNs predict equipment failures, optimize production schedules, and enhance overall system efficiency. For instance, a digital twin of a manufacturing plant equipped with ML techniques can analyze sensor data within a network to predict maintenance needs, reducing downtime and optimizing production output. In smart cities, ML-enabled DTNs contribute to the efficient management of smart city network infrastructure. For instance, in traffic management, digital twins simulate the behavior of the entire transportation network. ML algorithms analyze traffic patterns, predict congestion, and dynamically adjust traffic signals to optimize traffic flow. This results in reduced commute times and improved overall urban mobility. Besides, ML-driven DTNs can be also employed in energy grids to optimize power distribution over wireline networks. By creating a digital replica of the energy network infrastructure, ML algorithms can predict energy demand, identify potential faults, and optimize energy distribution in real time within the smart grid networks. This enhances the reliability and efficiency of the power grid, ensuring a stable energy supply.

In light of the coexistence of benefits and aforementioned challenges within such DT-inspired networking, this chapter aims to talk about the introduction first through providing the benefit evidence for using ML for DT creation and deployment. Then, this chapter provides the details that ML and DT can help in resource allocation, channel estimation, and user behavior prediction. Several aspects of reinforcement learning models are introduced within DT-enabled network applications. Lastly, the key insights of DTN is concluded.

3.2 ML for Creation and Synchronization of Digital Twin Network

This section focuses on the creation and synchronization of digital twin networks (DTNs), i.e., how to twin physical networks into DTNs, including (1) the state-of-the-art learning-based methods for mapping physical objects, network attributes, and protocols; and (2) the approaches to evolve DTNs that synchronize with the physical wireless networks taking into account their dynamic behaviors and twining efficiency for highly efficient operations.

3.2.1 Building Digital Twin Network

In order for DTN to succeed in nextG wireless networks, the physical network must be accurately mapped into a digital space with details, including both environmental layout and network attributes such as protocols, signal characteristics, relationship among communication entities, node mobility patterns, and application data usage.

In [7], several categories of digital twins are discussed, and these include product digital twins [8, 9], production digital twins [10, 11], and performance digital twins [12–14]. Although the above works have been taking off in many industries such as smart manufacturing, oil and gas, construction, and bio-engineering, digital twins remain relatively new for mobile and wireless networks. Particularly, [15–20] discussed how digital twin could be a powerful tool to fulfill the potential of 5G networks and beyond, which provided some insights on architectural trends, enabler techniques, and user cases, but not touching the core components of mapping or prototyping approaches. In practice, representing a physical network using a virtual model faces many challenges, especially to model physical objects, phenomenon, and network attributes *synergistically* into twins. Several works studied or partially addressed the creation and mapping problem of DTNs [21–23], which primarily adopted mathematical derivation and 3D modeling to twin the deployed objects in the environment, but struggled to map the physical wireless phenomenon that are close to real-time scenario. To address above limitations, [24–27] used data-driven ML-based approaches to effectively model the physical objects and wireless characteristics.

In essence, building DTNs that reflect the wireless networks will need to evolve significantly to address the technology and application trends such as high dynamics and mobility, high- frequency (mmWave/THz) communication, dense small cells, and virtualization. Instead of cloning all details of the physical network into DTNs directly, which is inefficient and even impractical, one promising approach is to map some essential environment information from the physical network space, and then create the corresponding network attributes in digital network twin through an environment-aware mapping approach, as discussed in [27], striking the balance between twinning efficiency and DTN accuracy. The underlying principle behind this idea is that several of these attributes are fully determined by the physical network space (PNS) which requires transferring to the digital space for creation of a DTN, while others could be created locally in DTN based on existing mapped components and contextual information. Specifically, a two-stage twinning method can be explained as follows (see Fig. 3.1):

(1) Twinning Data Generation In order to generate a large amount of representative data that capture both environment details and coupled network attributes for network twinning, one practical method is to synthetically generate high-quality data covering a wide range of fine-grained wireless network scenarios, which is then used to develop a faithful, efficient approach for the creation of network twins. In the light of the characteristic of highly environmental dependency in wireless networks, ray-tracing analysis becomes a powerful tool to generate a large amount of high-quality synthetic data given the environment details. Some existing works on quasi-deterministic (Q-D) ray tracers [27, 28] can be leveraged and extended to cover many wireless network scenarios of interests while complying with various standards and protocols, such as 3GPP LTE/NR and IEEE 802.11 ax/be/ay. To be more specific, the twinning dataset generation (TDG) framework based on ray-tracing analysis can be designed to randomly generate various instance data within a known scenario (i.e., physical network) covering different access point (AP)/user locations and protocol settings. Then, the TDG framework feeds all generated scenario cases into the Q-D ray tracer, which is used to capture the geometrical properties of the channel for each transceiver and generate the profile of delay τ, path gain, angle of departure (AoD) θ_t, angle of arrival (AOA) θ_r, etc, for the path components in the scenario. Note that any small change in the location of a node will be translated into changes in these captured profiles. Next, the output results from the ray tracer are directly used as input to the Q-D network channel model. Specifically, the channel impulse response of the Q-D channel can be characterized as [27]:

$$\begin{aligned} h(t) &= \sum_{\tau}\sum_{\theta_t}\sum_{\theta_r} Y_{tx}(\theta_t) \cdot Y_{rx}(\theta_r) \cdot h(t, \tau, \theta_t, \theta_r) \\ &= \sum_{i=0}^{N-1} 10^{-PL_i/20} e^{j\phi_i} \cdot (Y_{rx_i} \cdot Y_{tx_i}) \cdot e^{-j2\pi f \tau_i}, \end{aligned} \tag{3.1}$$

where N is the number of generated rays from ray tracer, PL_i and ϕ_i are the path loss and phase shift of ray i, and Y_{tx_i} and Y_{rx_i} are the radiation pattern of the transmitter and receiver array at ray i, respectively. For example, a power spectral representation of a mmWave frequency signal can be implemented in dividing the entire channel into a number of equally spaced sub-bands, and each of them has the specific size of corresponding to the sub-carrier spacing for an orthogonal frequency division multiplexing (OFDM) PHY, or the power is divided equally across all the sub-bands over the entire bandwidth for the single-carrier (SC) PHY mode. With the input of the Q-D trace files from the ray tracer, the TDG framework can parse these path profiles to obtain the spatial matrix between every transceiver pair. Specifically, the received power per sub-band Rx_i is computed and turned into a scalar value to represent the total energy apparent to the receiver by applying RF filtering as in [29], thus the overall received power is obtained by accumulating Rx_i over all sub-bands, and performance metrics, such as signal strength, signal-to-noise ratio (SNR), or packet throughput can further derived for any node distributions in the scenario. Note that, a few amount of measurement data from experimental testbeds will also contribute this data generation process in two aspects: (1) help derive the accurate and matched scenario distributions such as material reflectivity of objects; and (2) create a feedback path to tune the related Q-D channel parameters such that TDG framework can produce more accurate dataset that matches the actual measurements.

(2) Environment-Aware Attribute Prediction The communication characteristics over wireless networks typically depending on not only the relative positions of the nodes but also the environmental effects. To map the network attributes into DTNs, e.g., radio signal strength (RSS),[1] at arbitrary locations and time instances, by leveraging the environment details, such as surrounding objects' layout, sizes, material reflectivities, etc. To this end, data-driven methods using ML become the promising candidate to derive the mapping module by utilizing the data generated from the TDG framework, such as in [27]. For instance, an environment-aware attribute prediction (EAP) module can be trained that integrates propagation channel, node antenna configuration, tranceiver's power, and environment setups that produce a *radio map* containing RSS at any locations with respect to the transmitter, e.g., an access point (AP)/base station (BS). This learning-based model is especially suitable for nextG mmWave/THz wireless networks, where the directivity of antennas always impact the signal profiles and link-level performance. To train an EAP module for attribute mapping of DTN, one approach is to use a convolutional neural network (CNN) to analyze the data from the TDG framework for network attribute predictions. However, this direct application for digital twinning cannot capture time-varying signal dynamics since CNNs are primarily used for spatial data analysis. Thus, combining the model with long short-term memory (LSTM) and meta learning, such as in many spatial-temporal learning frameworks [30], becomes a potential solution, which will allow the model to analyze and capture both temporal and spatial features of physical network environments via only a few-shot training. For example, the physical environment factors and radio frequency parameters can be first fed into a CNN, which will extract the fine-grained features from the physical networks. The output of the CNN will be considered as an input of the LSTM that uses the current input with its recorded historical spatial features

[1] In what follows, this chapter takes the RSS as an example while such an approach can also be applicable for other network attributes.

of dynamic physical environments to predict future network attribute in DTN locally. Another potential solution is to use reinforcement learning (RL) to capture time-varying signal dynamics. RL serves as a powerful tool in predicting network dynamics by leveraging adaptive, self-learning algorithms. In the context of DTNs, RL can be applied to optimize attribute update processes in dynamic and complex environments. Through continuous interactions with the physical network system, RL models learn from the consequences of their actions, enabling them to adapt and improve attribute prediction over time. This approach is particularly valuable in predicting signal dynamics, as RL algorithms can capture communication patterns, adapt to changing network conditions, and optimize link quality allocation.

Such an integration of the synthetic data generation and environment-aware prediction makes it possible to efficiently map both physical environments and network attributes into digital network twins, covering a complete set of network attributes, e.g., radio maps, in both spatial and temporal domains. The feasibility of this DTN creation method can be implicitly demonstrated in a recent research work [27], which focused on predicting link quality at any locations of interest by leveraging environment details and physical object information. This work collected measurement and synthetic data and then used then to train a regression-based deep neural network (DNN) to predict a set of radio maps. These radio maps show the potential to assist in new performance evaluations of a network scenario with mobile users in digital network twins. In practical, constructing all radio maps using ray tracing or measurement would require several years of computation time for the given physical scenario, but combining ML technique with the ray-tracing calculation only takes several minutes to generate the complete set of radio maps. These results demonstrated the feasibility and potential of efficiently creating DTNs with the subsequent introduced mapping approaches, and also proved that the wireless network attributes can be discovered from environmental and physical object information.

3.2.2 Synchronizing Digital Twin Network and Physical Network

Before leveraging DTN in wireless network for various applications, it is crucial to periodically update it for the sake of keeping up with changes in the physical networks as illustrated in Fig. 3.1. Creating a DTN is a long-term process that can require significant resources depending on the complexity, scale and dynamics of the physical counterpart. This is especially true in a wireless network, where the wide range of physical objects, such as IoT sensors and mobile users moving around in the scenario, resulting in uncertainties in the channel due to fading and interference, which further complicate the task of synchronization between the physical network and its DTNs. In this regard, the second stage of building DTNs is to ensure that the mapped model remains relevant with periodic updates performed.

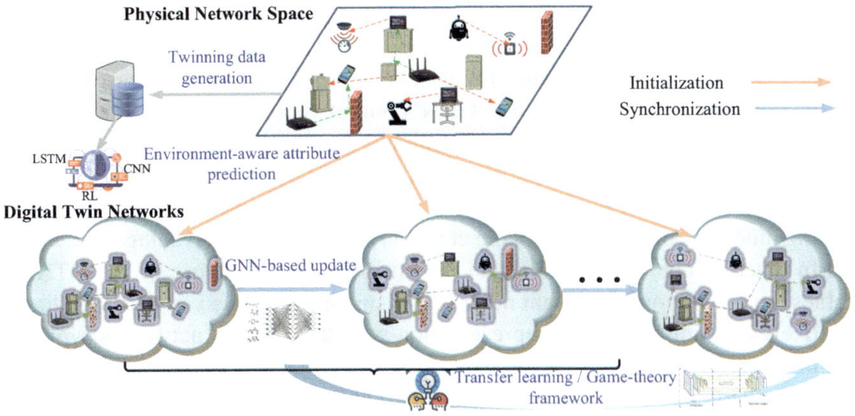

Fig. 3.1 DTN creation and synchronization with different ML models

Different from the direct mapping approach introduced in Sec. A, which gathers the information from physical environments to build a network twin, the goal of this twinning stage is to evolve the build DTN through the synchronization with the physical network dynamics. In a wireless environment, prior works such as [31, 32] have showed that when the physical network environment experiences a major change (i.e., the change of obstacle layout or AP/user locations), the network attributes could change accordingly, making it necessary to update the DTN that reflects the evolved network characteristic. To this end, the graphic neural network (GNN) is a natural solution to help evolve the existing DTN into a new twin that deals with the environment dynamics. Particularly, GNN [33] is a robust tool to analyze large-scale, complex systems with multi-faceted interdependencies, thus the GNN can be considered as a recipe to detect the variations of the physical network and update the DTN accordingly. To be more specific, the elements within the tangible network can initially be conceptualized as a graph structure. In this representation, tangible entities are designated as nodes, while the connections for communication are established as edges connecting various nodes. The condition of the communication channels and the state of the network can be quantified as weight values assigned to these edges. Within this framework, there are two distinct categories of nodes that emerge: (a) nodes equipped with communication capabilities, such as smartphones and Internet of Things (IoT) devices, and (b) nodes lacking communication capabilities, such as buildings and pieces of furniture. The first type of nodes possesses attributes encompassing both physical characteristics (such as location and size) and network-related properties. Conversely, the second type of nodes only possesses physical attributes.

Given these definitions, the utilization of GNNs is poised to play a crucial role in comprehending the evolving dynamics within the tangible space and subsequently reflecting these changes onto the existing DTN. In particular, the input for the GNN involves sampled feature vectors of neighboring nodes within one or two hops, encompassing their respective quantifiable characteristics. The GNN's output manifests as a probability vector denoted as $\boldsymbol{p} = [p_1, \ldots, p_N]$, where N symbolizes the number of considered attributes for each node. For instance, in the case of nodes lacking communication abilities, the GNN's output vector could be represented as $\boldsymbol{p} = [p_1, p_2]$, where p_1 signifies the likelihood of a change in the node's position and p_2 signifies the likelihood of a change in the node's dimensions. In conjunction with the GNN's output, the integration of an unsupervised network representation learning model, as proposed by Zhang et al. [34], could also be advantageous, which aids in determining whether to retain the attributes and structure of the DTN. It enables the adaptation of relevant nodes based on the collective vertex embeddings, effectively aligning the digital representation with the wireless network's evolving dynamics.

The fundamental principle underlying the application of GNN learning within the evolution of the DTN is rooted in the notion that the digital entities will mirror the same graph-based patterns exhibited by the tangible network. This alignment is achieved by progressively reflecting changes within the existing DTN. Moreover, the application of continuous learning (also known as lifelong learning), as suggested by Maltoni and Lomonaco [35], emerges as an optional approach to enhancing the synchronization process over time. This continuous learning approach aptly suits the dual evolution challenge, as each digital entity must steadily adapt using an ongoing influx of data harvested from the dynamic wireless network.

From the architectural perspective, the decentralization of DTN deployment is necessary to coordinate autonomous Internet of Everything (IoE) applications that exist in the real world (e.g., mobile devices and autonomous vehicles) with their respective digital counterparts in several sub-DTNs [36]. Thus, it is vital for such real-time DTNs to replicate the functionalities of the underlying physical applications while meeting the DTN synchronization requirements. Recent works have studied merging DTs with mobile edge computing (MEC) in an effort to meet synchronization demands, and assisted by ML approaches and game theory, several potential approaches for DTN synchronizations are:

(1) *Transfer learning coupled with deep reinforcement learning (DRL) approach* can be used to solve the DT-MEC placement and resource migration problems while minimizing the DT synchronization delay. Specifically, reinforcement learning (RL) is one of the widely adopted approaches to address dynamic programming problems. The efficiency of the traditional RL-based methods is relatively low because they require calculating the value functions of all possible state and action space pairs. The DRL explores policy and value functions through deep neural networks (DNN), which is considered as the most effective method to solve complex Markov decision process models. On the other hand, the transfer learning can allow the DTN to initialize a DRL model with weights learned from a previously initialized model (see Sec. I.A) on a related task or dataset. This initialization provides a good starting point, enabling the DTN to converge faster and require fewer training iterations to achieve good accuracy that tracks the physical network dynamics. On the other hand, such a learning model brings adaptation to domain shifts, for example, when the distribution of synchronization data in the current temporal domain differs from

the original domain, transfer learning can help the DRL model adapt DTN to these domain shifts more effectively. It allows the model to align its digital representations and policies with the characteristics of the evolved wireless network environment.

(2) *Game-theoretic framework with semantic extraction algorithms* [37] can be exploited in the resource allocation problem for wireless sensing devices with the goal of synchronizing the central DTN by controlling its sub-DTs' synchronization. On one hand, game theory is a mathematical framework used to study interactions and decision-making among multiple twins (players) in strategic situations. In the context of multi-DTN scenario, a game-theoretic framework is likely being employed to model the interactions and decision-making of the physical wireless devices, each acting as a player in a game. This framework could help in determining the optimal strategies for these devices to achieve synchronization and efficient resource allocation for twinning. On the other hand, semantic extraction algorithms refer to methods that extract meaningful information and context from data. In the context of DTN synchronization, these algorithms might be used to extract relevant information from the data collected from the physical wireless devices, such as environmental conditions, network status, or other relevant communication parameters. This extracted semantic information naturally minimizes the size of the transmitted data in an attempt to achieve enhanced real-time synchronization between physical networks and DTNs.

3.3 Digital Twin Networks Over Wireless Communications

Based on the methodology of building DTN from Sec. 3.2, this section introduces how to deploy and optimize these DTNs over actual wireless communication scenarios. The interplay between DT and wireless communications is essential for enabling the seamless integration of digital twins into various industries and applications, optimizing resource usage, ensuring data security, and meeting the diverse communication needs of network devices and systems. It is a multidisciplinary area that combines concepts from wireless networking, optimization theory, cyber-physical systems, and data analytics.

Consider the DTNs that consist of a PN and its mapping DTN generated and controlled by a central server. The total number of physical devices in the PN is denoted by K. The physical devices such as base stations and sensors need to transmit status data to the central server, which utilizes the obtained data to generate a DTN. Considering a long time period T and the time can be divided into N time slots, the duration of each time is then calculated as $T_0 = \frac{T}{N}$.

The channel gain between physical device k and the server at time slot n is denoted by h_{nk}. At time slot n, the physical device k in PN will generate data D_{nk}. To guarantee synchronization between the PN and the DTN, the physical device k can choose to transmit data d_{nk} to the server. The transmission rate between the physical device k and the server is given by

$$r_{nk} = B \log_2 \left(1 + \frac{p_{nk} h_{nk}}{\sigma^2}\right), \tag{3.2}$$

where B is the bandwidth of the system, p_{nk} is the transmit power of the physical device k at time slot n, and σ^2 is the noise power. Considering the randomness of the wireless channel, the received data at the server can contain error. The error rate of the transmission at time slot n can be given by Chen et al. [38]

$$\epsilon_{nk}(p_{nk}) = 1 - e^{-\frac{m\sigma^2}{p_{nk} h_{nk}}}, \tag{3.3}$$

with m being a waterfall threshold.

Since the transmitted data cannot be large than that of the remaining data, we have

$$\sum_{i=1}^{n} d_{ik} c_{ik}(p_{ik}) \leq \sum_{i=1}^{n} D_{ik}, \quad \forall n \in \mathcal{N}, k \in \mathcal{K}, \tag{3.4}$$

where

$$c_{ik}(p_{ik}) = \begin{cases} 1 \text{ with probability } e^{-\frac{m\sigma^2}{p_{ik} h_{ik}}} \\ 0 \text{ with probability } 1 - e^{-\frac{m\sigma^2}{p_{ik} h_{ik}}} \end{cases}, \tag{3.5}$$

$\mathcal{N} = \{1, \cdots, N\}$, and $\mathcal{K} = \{1, \cdots, K\}$.

In each time slot, let $x_{nk} \in \{0, 1\}$ denote whether physical device k in the PN transmits data to the server. The notation $x_{nk} = 1$ implies that physical device k transmits data to the BS; otherwise we have $x_{nk} = 0$. In this paper, frequency division multiple access (FDMA) is considered for the uplink transmission. Due to limited resource blocks of the communication system, the maximum number of associated users at each time slot is limited, i.e.,

$$\sum_{k=1}^{K} x_{nk} \leq K_0, \quad \forall n \in \mathcal{N}, \tag{3.6}$$

where K_0 is the maximum number of available resource blocks for the communication system.

At each time slot n, the wireless transmission delay and energy can be derived as

$$t_{nk} = \frac{d_{nk}}{r_{nk}}. \tag{3.7}$$

Since multiple users can simultaneously communicate with the server, the transmission delay of the PN at time slot n can be given by

$$t_n = \max_{k \in \mathcal{K}} x_{nk} t_{nk}. \tag{3.8}$$

Moreover, at each time slot n, the wireless transmission energy can be given by

$$e_{nk} = t_{nk} p_{nk}. \tag{3.9}$$

The accuracy of the DTN requires more data from the PN, while large data can lead to high transmission delay and energy. Thus, it is of importance to investigate the tradeoff between the accuracy and the wireless cost of the PN including delay and energy. The accuracy of each time slot n can be modeled as

$$a_n = f\left(\sum_{i=1}^{n} \sum_{k=1}^{K} d_{ik} c_{ik}(p_{ik}), \sum_{i=1}^{n} \sum_{k=1}^{K} D_{ik} \right), \tag{3.10}$$

where function $f(x, y) \in [0, 1]$ and $f(x, y)$ increases with x while decreases with y. The function of $f(x, y)$ can be obtained through simulations such as [39]. For example, we can set $f(x, y) = \left(\frac{x}{y}\right)^{\alpha}$, where $\alpha > 0$.

3.3.1 Problem Formulation

With the considered system model, our aim is minimize the transmission delay of the system with both accuracy and energy constraints. Mathematically, the optimization problem can be formulated as:

$$\min_{\boldsymbol{x}, \boldsymbol{d}, \boldsymbol{p}} \quad \sum_{n=1}^{N} \max_{k \in \mathcal{K}} x_{nk} t_{nk} \tag{3.11}$$

$$\text{s.t.} \quad f\left(\sum_{i=1}^{n} \sum_{k=1}^{K} d_{ik} c_{ik}(p_{ik}), \sum_{i=1}^{n} \sum_{k=1}^{K} D_{ik} \right) \geq A_n, \tag{3.11a}$$

$$\forall n \in \mathcal{N}, \tag{3.11b}$$

$$\sum_{n=1}^{N} t_{nk} p_{nk} \leq Q_k, \quad \forall k \in \mathcal{K}, \tag{3.11c}$$

$$\sum_{i=1}^{n} d_{ik} c_{ik}(p_{ik}) \leq \sum_{i=1}^{n} D_{ik}, \quad \forall n \in \mathcal{N}, k \in \mathcal{K}, \tag{3.11d}$$

$$\sum_{i=n}^{n+\tau} x_{ik} \geq \beta_k \tau, \quad \forall n \in \mathcal{N}, k \in \mathcal{K}, \tag{3.11e}$$

$$\sum_{k=1}^{K} x_{nk} \leq K_0, \quad \forall n \in \mathcal{N}, \tag{3.11f}$$

$$d_{nk} \geq 0, \quad \forall n \in \mathcal{N}, k \in \mathcal{K}, \tag{3.11g}$$

$$x_{nk} \in \{0, 1\}, \quad \forall n \in \mathcal{N}, k \in \mathcal{K}, \tag{3.11h}$$

$$0 \leq p_{nk} \leq P_k, \quad \forall n \in \mathcal{N}, k \in \mathcal{K}, \tag{3.11i}$$

$$0 \leq t_{nk} \leq T_0, \quad \forall n \in \mathcal{N}, k \in \mathcal{K}, \tag{3.11j}$$

where $\boldsymbol{x} = [x_{11}, \cdots, x_{1K}, \cdots, x_{NK}]^T$, $\boldsymbol{d} = [d_{11}, \cdots, d_{1K}, \cdots, d_{NK}]^T$, $\boldsymbol{p} = [p_{11}, \cdots, p_{1K}, \cdots, p_{NK}]^T$, Q_k is the maximum energy of physical device k, $\beta_k \in (0, 1]$ is a parameter to ensure that the physical device k and the server should have regular communication, τ is a constant parameter to ensure that the each physical device and the server must have at least one communication in τ time slots, and P_k is the maximum transmit power of physical device k.

Problem (3.11) is mixed integer optimization problem, which is generally hard to solve due to the following three difficulties. The first difficulty is the complicated accuracy function (3.10), of which the explicit expression is hard to obtain. The second difficulty lies in the nonconvex objective function (3.11) and constraints (3.11a)–(3.11c). The third difficulty lies in the integer scheduling variable x_{nk}. This chapter presents the proposed alternating algorithm to solve problem (3.11), which alternative solves three subproblems at each iteration., i.e., device scheduling subproblem, power control subproblem, and data offloading subproblem.

3.3.2 Device Scheduling Subproblem

With given transmission power and data offloading variables in problem (3.11), the device scheduling subproblem can be given by

$$\min_{\boldsymbol{x}} \quad \sum_{n=1}^{N} \max_{k \in \mathcal{K}} x_{nk} t_{nk} \tag{3.12}$$

$$\text{s.t.} \quad \sum_{i=n}^{n+\tau} x_{ik} \geq \beta_k \tau, \quad \forall n \in \mathcal{N}, k \in \mathcal{K}, \tag{3.12a}$$

$$\sum_{k=1}^{K} x_{nk} \leq K_0, \quad \forall n \in \mathcal{N}, \tag{3.12b}$$

$$x_{nk} \in \{0, 1\}, \quad \forall n \in \mathcal{N}. \tag{3.12c}$$

Problem (3.12) is a linear integer optimization problem. Relaxing the integer constraints and introducing slack variables y, Problem (3.12) becomes,

$$\min_{\boldsymbol{x},\boldsymbol{y}} \quad \sum_{n=1}^{N} y_n \tag{3.13}$$

$$\text{s.t.} \quad \sum_{i=n}^{n+\tau} x_{ik} \geq \beta_k \tau, \quad \forall n \in \mathcal{N}, k \in \mathcal{K}, \tag{3.13a}$$

$$\sum_{k=1}^{K} x_{nk} \leq K_0, \quad \forall n \in \mathcal{N}, \tag{3.13b}$$

$$y_n \geq x_{nk} t_{nk}, \quad \forall n \in \mathcal{N}, k \in \mathcal{K}, \tag{3.13c}$$

$$x_{nk} \in \{0, 1\} \quad \forall n \in \mathcal{N}, \tag{3.13d}$$

where $\boldsymbol{y} = [y_1, \cdots, y_N]^T$. Problem (3.13) is a linear integer optimization problem, which is generally hard to obtain the globally optimal solution. In the following, the dual method is used to solve the relaxed problem of (3.13) with replacing constraint (3.13d) with continuous constraint $x_{nk} \in [0, 1]$. It can be also proved that the obtained solution of the relaxed problem automatically satisfies the integer constraint in (3.13d). Through replacing constraint (3.13d) with continuous constraint $x_{nk} \in [0, 1]$, problem (3.13) can be transformed into

$$\min_{\boldsymbol{x},\boldsymbol{y}} \quad \sum_{n=1}^{N} y_n \tag{3.14}$$

$$\text{s.t.} \quad \sum_{i=n}^{n+\tau} x_{ik} \geq \beta_k \tau, \quad \forall n \in \mathcal{N}, k \in \mathcal{K}, \tag{3.14a}$$

$$\sum_{k=1}^{K} x_{nk} \leq K_0, \quad \forall n \in \mathcal{N}, \tag{3.14b}$$

$$y_n \geq x_{nk} t_{nk}, \quad \forall n \in \mathcal{N}, k \in \mathcal{K}, \tag{3.14c}$$

$$x_{nk} \in [0, 1] \quad \forall n \in \mathcal{N}. \tag{3.14d}$$

Since both objective function and constraints of problem (3.14) is convex, problem (3.14) is a convex problem and the optimal solution can be obtained by using the dual method. To obtain the optimal solution of problem (3.14), this section provides the following lemma.

Lemma 3.1 *The optimal solution of problem* (3.14) *is*

$$x^*_{nk} = \begin{cases} 1, \text{ if } \lambda_{2n} + \lambda_{3nk} - \sum_{i=\max\{n-\tau,1\}}^{n} \lambda_{1ik} < 0 \\ 0, \text{ else} \end{cases}, \tag{3.15}$$

and

$$y^*_n = \max_{k \in \mathcal{K}} x^*_{nk} t_{nk}, \tag{3.16}$$

where $\lambda_{1nk} \geq 0$, $\lambda_{2n} \geq 0$, *and* $0 \leq \lambda_{3nk} \leq 1$ *are the corresponding dual variables associated with constraints (3.14a)-(3.14c).*

Proof The dual function of problem (3.14) can be formulated by

$$\begin{aligned} &\mathcal{L}_1(\boldsymbol{x}, \boldsymbol{y}, \boldsymbol{\lambda}_1, \boldsymbol{\lambda}_2, \boldsymbol{\lambda}_3) \\ =& \sum_{n=1}^{N} y_n + \sum_{n=1}^{N} \sum_{k=1}^{K} \lambda_{1nk} \left(\beta_k \tau - \sum_{i=n}^{n+\tau} x_{ik} \right) + \sum_{k=1}^{K} \lambda_{2n} \left(\sum_{k=1}^{K} x_{nk} - K_0 \right) \\ &+ \sum_{n=1}^{N} \sum_{k=1}^{K} \lambda_{3nk} \left(x_{nk} t_{nk} - y_n \right), \end{aligned}$$

where $\boldsymbol{\lambda}_1 = [\lambda_{111}, \cdots, \lambda_{1NK}]^T$, $\boldsymbol{\lambda}_2 = [\lambda_{21}, \cdots, \lambda_{2N}]^T$, and $\boldsymbol{\lambda}_3 = [\lambda_{311}, \cdots, \lambda_{3NK}]^T$. Considering constraint (3.14d) and $\mathcal{L}_1(\boldsymbol{x}, \boldsymbol{y}, \boldsymbol{\lambda}_1, \boldsymbol{\lambda}_2, \boldsymbol{\lambda}_3)$ is a linear function with respect to $\boldsymbol{x}$, the optimal solution of $\boldsymbol{x}$ can be derived as (3.15). Further considering that y_n only has lower bound as shown in (3.14c), the dual function $\mathcal{L}_1(\boldsymbol{x}, \boldsymbol{y}, \boldsymbol{\lambda}_1, \boldsymbol{\lambda}_2, \boldsymbol{\lambda}_3)$ is unbounded unless $1 - \lambda_{3nk} \geq 0$. Thus, we always have $1 - \lambda_{3nk} \geq 0$ and the optimal solution of y^*_n should be the minimum value satisfying (3.14c), as shown in (3.16). □

According to Lemma 3.1, the optimal solution of $\boldsymbol{x}$ and $\boldsymbol{y}$ can be obtained with given dual variables. Besides, the value of x_{nk} is always zero or one, which automatically satisfies the integer constraint (3.13d). With obtained $\boldsymbol{x}$ and $\boldsymbol{y}$, the values of the dual variables can be updated with the sub-gradient method [40]:

$$\lambda_{1nk}(t+1) = \left[\lambda_{1nk}(t) + \kappa(t) \left(\beta_k \tau - \sum_{i=n}^{n+\tau} x_{ik} \right) \right]^+, \tag{3.17}$$

$$\lambda_{2n}(t+1) = \left[\lambda_{2n}(t) + \kappa(t) \left(\sum_{k=1}^{K} x_{nk} - K_0 \right) \right]^+, \tag{3.18}$$

$$\lambda_{3nk}(t+1) = \left(\lambda_{3nk}(t) + \kappa(t) \left(x_{nk} t_{nk} - y_n \right) \right) |_0^1, \tag{3.19}$$

where (t) denotes the value of variable in the t-th iteration, $\kappa(t)$ is the dynamatic stepsize of the dual method, $[x]^+ = \max\{x, 0\}$, and $x_a^b = \max\{\min\{x, a\}, b\}$. Through iteratively updating the value of $\boldsymbol{x}$ and $\boldsymbol{y}$ with the dual variables, the optimal solution is obtained.

3.3.3 Power Control Subproblem

With given device scheduling and data offloading variables in problem (3.11), the power control subproblem can be formulated as

$$\min_{\boldsymbol{p}} \quad \sum_{n=1}^{N} \max_{k\in\mathcal{K}} x_{nk} t_{nk} \tag{3.20}$$

$$\text{s.t.} \quad a_n \geq A_n, \quad \forall n \in \mathcal{N}, \tag{3.20a}$$

$$\sum_{n=1}^{N} t_{nk} p_{nk} \leq Q_k, \quad \forall k \in \mathcal{K}, \tag{3.20b}$$

$$\sum_{i=1}^{n} d_{ik} c_{ik}(p_{ik}) \leq \sum_{i=1}^{n} D_{ik}, \quad \forall n \in \mathcal{N}, k \in \mathcal{K}, \tag{3.20c}$$

$$0 \leq p_{nk} \leq P_k, \quad \forall n \in \mathcal{N}, \tag{3.20d}$$

$$0 \leq t_{nk} \leq T_0, \quad \forall n \in \mathcal{N}. \tag{3.20e}$$

Problem (3.20) is hard to solve due to the complicated constraints (3.20a)–(3.20c). In order to handle constraint (3.20c), the expect value of $c_{ik}(p_{ik})$ is used to represent the value of $c_{ik}(p_{ik})$, i.e., we have

$$c_{ik}(p_{ik}) = e^{-\frac{m\sigma^2}{p_{ik} h_{ik}}}. \tag{3.21}$$

Without loss of generality, this section considers the expression of accuracy function $f(x, y) = \left(\frac{x}{y}\right)^\alpha$ with α being the constant to be determined in the simulations in the following analysis. Further substituting the expressions of $t_{nk} = \frac{d_{nk}}{r_{nk}}$ and $r_{nk} = B \log_2 \left(1 + \frac{p_{nk} h_{nk}}{\sigma^2}\right)$ as well as introducing slack variables t_{nk} and z_n, problem (3.20) can be equivalent to

$$\min_{\boldsymbol{p},\boldsymbol{t},\boldsymbol{z}} \quad \sum_{n=1}^{N} z_n \tag{3.22}$$

$$\text{s.t.} \quad z_n \geq x_{nk} t_{nk}, \quad \forall n \in \mathcal{N}, k \in \mathcal{K}, \tag{3.22a}$$

$$t_{nk} \geq \frac{d_{nk}}{B\log_2\left(1+\frac{p_{nk}h_{nk}}{\sigma^2}\right)}, \quad \forall n \in \mathcal{N}, k \in \mathcal{K}, \tag{3.22b}$$

$$\sum_{i=1}^{n}\sum_{k=1}^{K} d_{ik} e^{-\frac{m\sigma^2}{p_{ik}h_{ik}}} \geq A_n^{1/\alpha} \sum_{i=1}^{n}\sum_{k=1}^{K} D_{nk}, \quad \forall n \in \mathcal{N}, \tag{3.22c}$$

$$\sum_{n=1}^{N} t_{nk} p_{nk} \leq Q_k, \quad \forall k \in \mathcal{K}, \tag{3.22d}$$

$$\sum_{i=1}^{n} d_{ik} e^{-\frac{m\sigma^2}{p_{ik}h_{ik}}} \leq \sum_{i=1}^{n} D_{ik}, \quad \forall n \in \mathcal{N}, k \in \mathcal{K}, \tag{3.22e}$$

$$0 \leq p_{nk} \leq P_k, \quad \forall n \in \mathcal{N}, \tag{3.22f}$$

$$0 \leq t_{nk} \leq T_0, \quad \forall n \in \mathcal{N}, \tag{3.22g}$$

where $\boldsymbol{z} = [z_1, \cdots, z_N]^T$. Problem (3.22) is nonconvex due to constraints (3.22b)-(3.22e). To handle the nonconvexity of (3.22b) and (3.22d), variable substitution can be used. Introducing a new variable $q_{nk} = t_{nk} p_{nk}$ and replacing p_{nk} with q_{nk}, problem (3.22) becomes

$$\min_{\boldsymbol{q},\boldsymbol{t},\boldsymbol{z}} \quad \sum_{n=1}^{N} z_n \tag{3.23}$$

$$\text{s.t.} \quad (3.22a),\ (3.22f),\ (3.22g)$$

$$t_{nk} B\log_2\left(1+\frac{q_{nk}h_{nk}}{\sigma^2 t_{nk}}\right) \geq d_{nk}, \quad \forall n \in \mathcal{N}, k \in \mathcal{K}, \tag{3.23a}$$

$$\sum_{i=1}^{n}\sum_{k=1}^{K} d_{ik} e^{-\frac{m\sigma^2 t_{ik}}{q_{ik}h_{ik}}} \geq A_n^{1/\alpha} \sum_{i=1}^{n}\sum_{k=1}^{K} D_{ik}, \quad \forall n \in \mathcal{N}, \tag{3.23b}$$

$$\sum_{n=1}^{N} q_{nk} \leq Q_k, \quad \forall k \in \mathcal{K}, \tag{3.23c}$$

$$\sum_{i=1}^{n} d_{ik} e^{-\frac{m\sigma^2 t_{ik}}{q_{ik}h_{ik}}} \leq \sum_{i=1}^{n} D_{ik}, \quad \forall n \in \mathcal{N}, k \in \mathcal{K}, \tag{3.23d}$$

where both constraints (3.23a) and (3.23c) are convex now. As a result, it remains to solve the complex exponential expressions in constraints (3.23b) and (3.23d).

In order to handle the non-convexity of constraints (3.23b) and (3.23d), the first-order Taylor series can be used to approximate $e^{-\frac{m\sigma^2 t_{ik}}{q_{ik}h_{ik}}}$, which can be given by

$$\sum_{i=1}^{n}\sum_{k=1}^{K} d_{ik} e^{-\frac{m\sigma^2 t_{ik}^{(m)}}{q_{ik}^{(m)} h_{ik}}} \left(1 - \frac{m\sigma^2}{q_{ik}^{(m)} h_{ik}}(t_{ik} - t_{ik}^{(m)}) + \frac{m\sigma^2 t_{ik}^{(m)}}{(q_{ik}^{(m)})^2 h_{ik}}(q_{ik} - q_{ik}^{(m)})\right) \geq A_n^{1/\alpha} \sum_{i=1}^{n}\sum_{k=1}^{K} D_{ik} \tag{3.24}$$

and

$$\sum_{i=1}^{n} d_{ik} e^{-\frac{m\sigma^2 t_{ik}^{(m)}}{q_{ik}^{(m)} h_{ik}}} \left(1 - \frac{m\sigma^2}{q_{ik}^{(m)} h_{ik}}(t_{ik} - t_{ik}^{(m)}) + \frac{m\sigma^2 t_{ik}^{(m)}}{(q_{ik}^{(m)})^2 h_{ik}}(q_{ik} - q_{ik}^{(m)})\right) \leq \sum_{i=1}^{n} D_{ik}, \tag{3.25}$$

where $q_{ik}^{(m)}$ and $t_{ik}^{(m)}$ are respectively the values of q_{ik} and t_{ik} in the m-th iteration. According to (3.24) and (3.25), the convex linear inequality constraints are obtained.

Through replacing constraints (3.23b) and (3.23d) with (3.24) and (3.25) respectively, problem (3.23) becomes the following convex problem:

$$\min_{\boldsymbol{q},\boldsymbol{t},\boldsymbol{z}} \sum_{n=1}^{N} z_n \tag{3.26}$$

s.t. (3.22a), (3.22f), (3.22g)

$$t_{nk} B \log_2\left(1 + \frac{q_{nk} h_{nk}}{\sigma^2 t_{nk}}\right) \geq d_{nk}, \quad \forall n \in \mathcal{N}, k \in \mathcal{K}, \tag{3.26a}$$

$$\sum_{i=1}^{n}\sum_{k=1}^{K} d_{ik} e^{-\frac{m\sigma^2 t_{ik}^{(m)}}{q_{ik}^{(m)} h_{ik}}} \left(1 - \frac{m\sigma^2}{q_{ik}^{(m)} h_{ik}}(t_{ik} - t_{ik}^{(m)}) + \frac{m\sigma^2 t_{ik}^{(m)}}{(q_{ik}^{(m)})^2 h_{ik}}(q_{ik} - q_{ik}^{(m)})\right)$$
$$\geq A_n^{1/\alpha} \sum_{i=1}^{n}\sum_{k=1}^{K} D_{ik}, \quad \forall n \in \mathcal{N}, \tag{3.26b}$$

$$\sum_{n=1}^{N} q_{nk} \leq Q_k, \quad \forall k \in \mathcal{K}, \tag{3.26c}$$

$$\sum_{i=1}^{n} d_{ik} e^{-\frac{m\sigma^2 t_{ik}^{(m)}}{q_{ik}^{(m)} h_{ik}}} \left(1 - \frac{m\sigma^2}{q_{ik}^{(m)} h_{ik}}(t_{ik} - t_{ik}^{(m)}) + \frac{m\sigma^2 t_{ik}^{(m)}}{(q_{ik}^{(m)})^2 h_{ik}}(q_{ik} - q_{ik}^{(m)})\right)$$
$$\leq \sum_{i=1}^{n} D_{ik}, , k \in \mathcal{K}, \tag{3.26d}$$

which can be effectively solved through using the existing convex optimization algorithms. Trough solving problem 3.26, the solution of the original problem (3.20) can be effectively obtained.

3.3.4 Data Offloading Subproblem

With given device scheduling and power control variables in problem (3.11), the data offloading subproblem can be rewritten as

$$\min_{\boldsymbol{d}} \quad \sum_{n=1}^{N} \max_{k \in \mathcal{K}} x_{nk} t_{nk} \tag{3.27}$$

$$\text{s.t.} \quad a_n \geq A_n, \quad \forall n \in \mathcal{N}, \tag{3.27a}$$

$$\sum_{n=1}^{N} t_{nk} p_{nk} \leq Q_k, \quad \forall k \in \mathcal{K}, \tag{3.27b}$$

$$\sum_{i=1}^{n} d_{ik} c_{ik}(p_{ik}) \leq \sum_{i=1}^{n} D_{ik}, \quad \forall n \in \mathcal{N}, k \in \mathcal{K}, \tag{3.27c}$$

$$d_{nk} \geq 0, \quad \forall n \in \mathcal{N}, \tag{3.27d}$$

$$0 \leq t_{nk} \leq T_0, n \in \mathcal{N}. \tag{3.27e}$$

Substituting (3.7) and (3.10) into problem (3.27) yields

$$\min_{\boldsymbol{d}} \quad \sum_{n=1}^{N} \max_{k \in \mathcal{K}} \frac{x_{nk} d_{nk}}{r_n k} \tag{3.28}$$

$$\text{s.t.} \quad \sum_{i=1}^{n} \sum_{k=1}^{K} d_{ik} e^{-\frac{m\sigma^2}{p_{ik} h_{ik}}} \geq A_n^{1/\alpha} \sum_{i=1}^{n} \sum_{k=1}^{K} D_{ik}, \quad \forall n \in \mathcal{N}, \tag{3.28a}$$

$$\sum_{n=1}^{N} \frac{p_{nk} d_{nk}}{r_n k} \leq Q_k, \quad \forall k \in \mathcal{K}, \tag{3.28b}$$

$$\sum_{i=1}^{n} d_{ik} c_{ik}(p_{ik}) \leq \sum_{i=1}^{n} D_{ik}, \quad \forall n \in \mathcal{N}, k \in \mathcal{K}, \tag{3.28c}$$

$$d_{nk} \geq 0, \quad \forall n \in \mathcal{N}, \tag{3.28d}$$

$$0 \leq t_{nk} \leq T_0, \quad n \in \mathcal{N}, \tag{3.28e}$$

which is a linear programming problem and can be effectively solved by using the simplex method.

Algorithm 1 Alternating algorithm

1: Initialize a feasible solution $(\boldsymbol{x}^{(0)}, \boldsymbol{d}^{(0)}, \boldsymbol{p}^{(0)})$ of problem (3.11) and set $l = 0$.
2: **repeat**
3: With given $(\boldsymbol{d}^{(l)}, \boldsymbol{p}^{(l)})$, obtain the solution $\boldsymbol{x}^{(l+1)}$ of problem (3.12).
4: With given $(\boldsymbol{x}^{(l+1)}, \boldsymbol{d}^{(l)})$, obtain the solution $\boldsymbol{p}^{(l+1)}$ of problem (3.20).
5: With given $(\boldsymbol{x}^{(l+1)}, \boldsymbol{p}^{(l+1)})$, obtain the solution $\boldsymbol{d}^{(l+1)}$ of problem (3.27).
6: Set $l = l + 1$.
7: **until** objective value (3.11) converges

3.3.5 Algorithm Analysis

Through alternatively solving subproblems (3.12), (3.20), and (3.27), the overall procedure to obtain a solution of problem (3.11) can be shown in Algorithm 1. Since the objective value (3.11) is nonincreasing and the objective value (3.11) has a limited lower bound (i.e., zero), Algorithm 1 always converges.

The main complexity of Algorithm 1 lies in solving subproblems (3.12), (3.20), and (3.27). To solve subproblem (3.12), the optimal solution can be obtained through Lemma 3.1 with complexity $\mathcal{O}(NK\tau)$ for given dual variables. Thus, the complexity of solving subproblem (3.12) is $\mathcal{O}(I_1 NK\tau)$, where I_1 is the total number of iterations of using the dual method. To solve subproblem (3.20), the total complexity is $\mathcal{O}(I_1 N^3 K^3)$ through solving a series of convex subproblems (3.26), where I_2 is the total number of iterations using the successive convex approximation method. To solve subproblem (3.27), the total complexity is $\mathcal{O}(N^{2.5} K^{2.5})$ through the simplex method. As a result, the total complexity of Algorithm 1 is $\mathcal{O}(I_0 I_1 N^3 K^3)$, where I_0 denotes the total number of outer iterations.

3.3.6 Simulation Results

This section provides the simulation results of the proposed algorithm. Consider a PN with K physical devices uniformly distributed in a square area with size 200×200 m. The path loss model is $128.1 + 37.6 \log_{10} d$ (d is in km) and the standard deviation of shadow fading is 8 dB [41]. The bandwidth is $B = 1$ MHz and the noise power spectral density is $N_0 = -174$ dBm/Hz. Unless otherwise specified, this section sets the number of physical devices $K = 10$, the total number of time slots $N = 10$, equal accuracy requirement $A_1 = \cdots = A_n = 0.6$, equal arriving data $D_{11} = \cdots = D_{NK} = 300$ kbits, parameter $\tau = 3$, $K_0 = 5$, equal constant $\beta_1 = \cdots = \beta_K = 1/3$, and equal maximum transmit power $P_1 = \cdots = P_K = 1$ dBm.

To show the effectiveness of the proposed algorithm, it is natural to consider the following two baselines: the random device selection algorithm (labeled as 'Random'), where the power control and data offloading are optimized by using the proposed algorithm, and the adaptive edge association algorithm in [42] (labeled

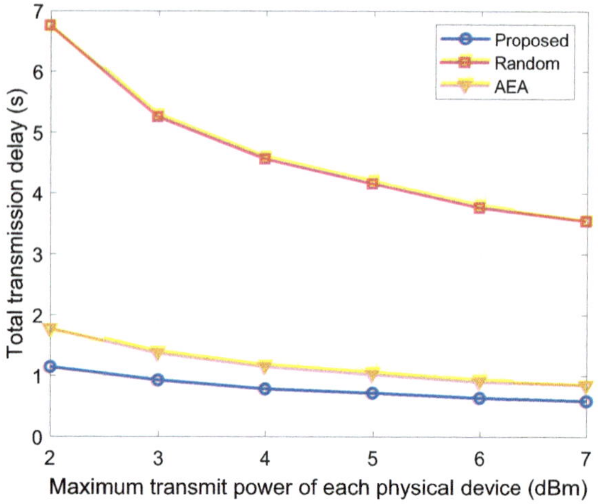

Fig. 3.2 The transmission delay versus the maximum transmit power of each physical device with $K = 14$

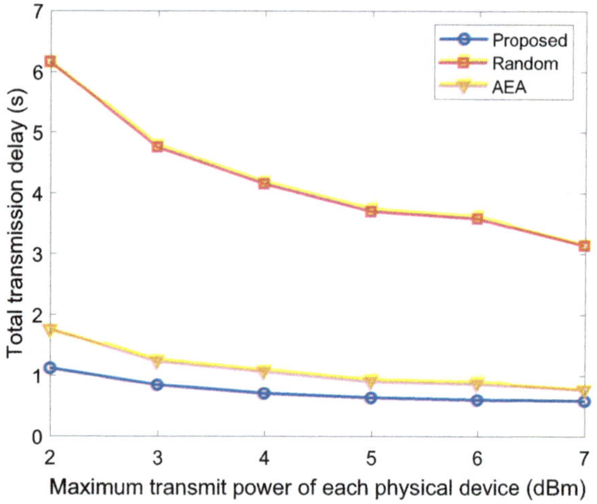

Fig. 3.3 The transmission delay versus the maximum transmit power of each physical device with $K = 10$

as 'AEA'), where the device selection and data offloading are solved by using the proposed algorithm.

In Figs. 3.2 and 3.3, it shoes the transmission delay versus the maximum transmit power of each physical device with various number of physical devices. From both figures, it can be observed that the total transmission delay of all algorithms decreases with the maximum transmit power and the decreasing speed is decreasing. The reason is that large maximum transmit power can allow the physical device to transmit with high power, thus reducing the transmission delay. Compared with AEA in [42], the proposed algorithm can yield lower transmission delay expecially when the maximum transmit power is small. The reason is that the proposed algorithm jointly optimizes multiple time slots, while the resource is not cooperatively scheduled in AEA.

3.4 Reinforcement Learning for Digital Twin-Assisted Wireless Networks

Given the creation and deployment of DT in wireless networks, DTN offers several advantages when combined with reinforcement learning (RL) for optimizing wireless networks. First, the built DTNs provide a controlled environment for testing and experimenting with various resource allocation strategies without affecting the real network, where RL algorithms can learn and adapt within this virtual environment, allowing for rapid prototyping and risk-free exploration of different approaches. Second, DTNs can generate large amounts of synthetic data that can be used for training RL agents over wireless networks. This synthetic data complements real-world data, allowing RL algorithms to learn from a diverse range of scenarios, including rare or extreme events that might not occur frequently in the actual network. Third, scaling up a real network for extensive experimentation can be impractical, but DTNs remove this limitation by facilitating large-scale experiments in the virtual environment, enabling exploration of complex interactions and network behavior when using RL for optimization. Last but not least, DTNs allow researchers to test RL algorithms over wireless networks without the need for access to sensitive real-world data. This is particularly valuable in cases where user privacy and data security are concerns.

3.4.1 Dynamic Resource Allocation Using Reinforcement Learning and Digital Twin Networks

To address the challenges of efficiently managing limited resources in dynamic and complex wireless communication environments, paring reinforcement learning techniques with DTNs has potential to adaptively optimize the allocation of resources in wireless networks. Here is a more detailed breakdown of each component in such an optimization problem:

(1) *Dynamic resource allocation*: Wireless networks encompass various resources like frequency bands, time slots, power levels, and bandwidth. These resources need to be allocated dynamically to meet the changing demands of users, devices, and applications. Traditional static resource allocation schemes allocate fixed resources to users regardless of changing network conditions, leading to inefficient resource usage and potential underutilization or overloading. In contrast, dynamic resource allocation continuously monitors the network environment, collects relevant data, and employs algorithms, such as reinforcement learning, to make informed decisions about resource allocation on the fly.
(2) *RL*: RL is a machine learning paradigm where an agent learns how to take actions in an environment to maximize a cumulative reward signal over time. In the context of wireless networks, RL algorithms can be employed to learn

optimal resource allocation policies. The agent interacts with the environment (the network) and learns which actions lead to better performance based on feedback from rewards. Specifically, a standard reinforcement learning problem can be generalized as a Markov decision process (MDP), defined by the tuple $\langle \mathcal{S}, \mathcal{A}, \gamma, r, P \rangle$, where $\mathcal{S}$ is the current state, $\mathcal{A}$ is the action spaces, $r : \mathcal{S} \times \mathcal{A} \rightarrow \mathbb{R}$ is the reward function, γ is the corresponding discounting factor, and $P(\cdot|s, a)$ is the state transition dynamics. The goal of this reinforcement learning algorithm is to find the optimal state-action pair to solve the resource allocation problem dynamically. Starting from an initial state, the agent selects an action at the t-th decision epoch and transitions from the current state to the next state according to a probability distribution. Simultaneously, the agent receives a reward to evaluate the chosen action. Consequently, the transition probability and reward are dependent on both states and the action taken.

The corresponding value function, denoted as $V^{\pi}(s)$, is a fundamental concept in reinforcement learning. It captures the expected cumulative discounted reward obtained by following a policy π and observing the environment dynamics. The value function is defined as the expected sum of discounted rewards, starting from state s_0, given by:

$$V^{\pi}(s) = \mathbb{E}\left[\sum_{t=0}^{\infty} \gamma^t \cdot R_t, \middle|, S_0 = s\right], \tag{3.29}$$

where γ represents the discount factor, R_t is the reward obtained at time t, and s denotes the initial state. To compute the action value function, denoted as $Q^{\pi}(s, a)$, the rewards, states, and transition probabilities are considered to associate with each resource allocation action. The action value function is defined as follows:

$$Q^{\pi}(s, a) = r(s, a) + \gamma \cdot \mathbb{E}\left[V^{\pi}(S'), \middle|, S' \sim P(\cdot \mid s, a)\right], \tag{3.30}$$

where $r(s, a)$ represents the immediate reward obtained by taking action a in state s, and S' follows the transition probability $P(\cdot|s, a)$. By employing the law of iterative expectation, one can express the value function $V^{\pi}(s)$ in terms of the action value function $Q^{\pi}(s, a)$ for any policy π. Such relationship can be formulated by:

$$V^{\pi}(s) = \mathbb{E}\left[Q^{\pi}(s, A), \middle|, A \sim \pi(\cdot \mid s)\right], \quad \forall s \in S, \tag{3.31}$$

where A represents the action chosen according to policy $\pi(\cdot \mid s)$. In order to reach optimality, a greedy policy π_Q can be defined to select actions based on the highest action values Q. The greedy policy π_Q is defined as the policy that selects actions based on the maximum action values Q. In other words, for a given state s, the greedy policy chooses the action a that maximizes the action value function $Q^{\pi}(s, a)$. By employing the greedy policy π_Q, one can continuously refine the value function estimates and improve the decision-making process toward achieving optimal behavior in a RL based resource allocation setting.

(3) *DTN-assisted solution*: A digital twin is a virtual representation of a physical network system or process. In the context of wireless networks, the DTN can simulate the behavior of the network, including interactions between devices, communication patterns, interference, and resource allocation dynamics. This simulation provides a realistic environment for testing and optimizing resource allocation strategies. Bringing the above components together, the approach involves the following steps:

(a) *Modeling the network as a digital twin*: Create the virtual model of the wireless network, capturing the various components, nodes, devices, user behaviors, and network conditions, as introduced in Sec. 3.2. Such a DTN simulates the network's behavior, responses, and interactions.

(b) *RL agent training in DTN*: Train an RL agent within the digital twin environment. DTN is exploited to monitor the model's training state and complex network environment, where the network state $s(t) \in S$ is created by the DTN and transmitted to the RL agent. Then, the agent learns by taking actions (resource allocation decisions) and receiving feedback (rewards) from the simulated network based on its performance. The goal is to learn resource allocation policies that optimize network objectives, such as throughput, latency, or energy efficiency.

(c) *Dynamic policy optimization*: The RL agent's learning process aims to discover resource allocation policies that adapt to changing network conditions and user demands. As the agent interacts with the DTN that covers plenty of situations and synchronizes with the physical network, it is able to continuously refines the policy to make on-demand resource allocation decisions.

(d) *Transfer to Real Network*: Once the RL agent has learned effective policies within the DTN, these policies can be transferred and tested in the real wireless network. The agent's learned policies should adapt to the real world dynamics observed, but the initial training in the DTN accelerates the process and reduces risks associated with direct real-world experimentation.

By combining reinforcement learning with DTNs, this approach enables efficient and effective dynamic resource allocation in wireless networks. In essence, the DTN provides a salable, controllable, and synchronized environment for learning and experimentation, while RL agents optimize resource allocation policies that can be applied to real-world scenarios.

3.4.2 Intelligent Edge Caching with Reinforcement Learning and Digital Twin Networks

The significance of caching in addressing the rising need for high-rate, low-latency services in the next generation (nextG) networks cannot be overstated. The act of storing frequently accessed content like videos, images, and applications in close proximity to users has a substantial impact. It notably reduces the time and resources needed for retrieving data, not only improving the user experience by minimizing delays but also easing the strain on the network's infrastructure, thus boosting its overall efficiency. Furthermore, caching plays a critical part in enabling the seamless provision of emerging data-intensive services such as augmented reality (AR), virtual reality (VR), video streaming, and autonomous vehicles, which are projected to become even more prevalent in the era of 5G and 6G [43–45].

However, implementing edge caching in wireless networks presents several challenges and safety considerations that require careful attention. One core challenge is the dynamic nature of user demands and the ever-changing popularity of content. This makes accurately predicting which data should be cached or removed, and determining the optimal storage cache space, quite challenging. Moreover, these cache replacement decisions need to prioritize the safety of the system by ensuring that base stations (BSs) remain balanced and do not get overwhelmed. For instance, if a particular BS is operating near its capacity limit, it becomes crucial to distribute or replace the relevant content among other BSs that have sufficient capacity. This ensures a well-balanced and optimized system, maintaining its efficiency and stability. Importantly, neglecting this could lead to severe consequences like system crashes (e.g., resource depletion and network congestion), posing a threat to the network's overall functioning and significantly degrading the user experience. Therefore, in the interest of network reliability and sustainability, it is imperative to implement robust safety mechanisms and algorithms in edge caching. These mechanisms should be capable of continuously monitoring environmental conditions, dynamically adapting caching strategies, and efficiently distributing the load among BSs. By underscoring the importance of system safety and upholding a well-balanced network infrastructure, the network can operate optimally, mitigating the risk of disruptions and ensuring uninterrupted service delivery.

With this objective in mind, the integration of safe reinforcement learning alongside DTNs emerges as a desirable approach for the advancement of intelligent edge caching. Safe RL can be augmented with several effective modules within DTNs, i.e., a measure taken to avert safety concerns such as the imbalance of load among BSs. The incorporation of safe RL algorithms facilitates astute decision-making by accommodating safety restrictions and avoiding potential risks. These algorithms acquire strategic insights from present data and interactions with the environment, ensuring the security of caching decisions while mitigating conceivable hazards. Concurrently, DTNs can assume roles as both RL optimizer and safeguards throughout the caching process. Through the emulation of network behavior and precise anticipation of the implications of caching choices, DTNs empower the optimization of caching tactics within a regulated setting. Furthermore, they serve as protective mechanisms by actively overseeing and validating the integrity of cached content, guaranteeing resilience against malware and other safety vulnerabilities. For instance, DTN can be harnessed to simulate infrequent and rare network scenarios that could arise in the future, thus promoting minimal risk and network breakdowns. To be specific, this intelligent edge caching strategy can be succinctly outlined in the following parts:

The developments in nextG cellular networks are poised to incorporate dense deployments of small-cell BSs. Consider a network comprising N BSs, all interlinked with a common data center through high-latency, low-bandwidth backhaul connections. Each BS is equipped with M cache lines capable of housing content sourced from the data center. For the sake of simplicity, one can assume that the content aligns with the size of a cache line, and during each time slot t, a specific count of contents is expected to be requested by end users. In specific scenarios where the requested content is already stored in the BS (referred to as a *cache hit*), the content can be swiftly retrieved. However, if it is not cached, the content must be procured from the data center through the backhaul links, inevitably incurring higher communication cost and additional transmission delays. The concept of cache replacement involves proactively fetching particular contents via the backhaul and pre-storing them within the BSs before user requests materialize. Hence, the core objectives of intelligent edge caching are two-fold: (1) maximize the average cache hit ratio and (2) minimize the peak load on the backhaul link, thereby enhancing operational efficiency.

(1) RL Model for Caching: The safe RL framework can be implemented based on Deep Q Network (DQN) [46], which is a representative RL algorithm to solve the above cache replacement problem. The RL algorithm runs on the a cloud server and monitors the running states of BSs. The state s is defined as a tuple of (n, t_n, f_n), where n is the BS index. t_n and f_n denote the last time at which the cache line cached and its caching frequency, respectively. Based on the state s, the policy model $\pi(\cdot|s)$ can be employed to determine for each BS *whether* to skip or update a cache line and, if so, *which* specific cache line to update. The action space is thus defined as $\{\epsilon, C_1, C_2, \ldots, C_M\}$, where ϵ denotes skipping caching and C_m corresponds to updating the m-th cache slot. Actions are only taken when a cache miss occurs, indicating that the requested data is not present in the cache.

The model is globally rewarded right after the caching action. Specifically, the reward function can be defined as:

$$r(s, a, s') = H(s, a, s') + \lambda \cdot P(s, a, s'), \tag{3.32}$$

where $H(s, a, s')$ and $P(s, a, s')$ denote the reward and penalty at time t, respectively. λ is a hyperparameter to balance the ratio between reward and penalty. Let $h(t)$ denote the cumulative number of cache hits until time t. $h(t+1) - h(t)$ represents the number of cache hits generated at time $t+1$. Therefore, the cache reward can be derived as follow:

$$H(s, a, s') = \alpha \cdot (h(t+1) - h(t)), \tag{3.33}$$

where α is a hyperparameter that weights the number of cache hits. The penalty term $P(s, a, s')$ punishes the case that the current cache miss is caused by previous actions, which can be defined as follows:

$$P(s, a, s') = \frac{1}{[H(s, a, s')]^{\kappa} + 1}, \tag{3.34}$$

where κ is a hyperparameter to control the significance of the penalty.

In the training of DQN, two key techniques are considered and fit well into this catching problem—experience replay and target network [47]. Experience replay involves the use of a memory buffer to store the agent's past caching experiences during the training process. These experiences are represented as tuples, $\langle \mathcal{S}, \mathcal{A}, r, \mathcal{S}' \rangle$, which are then randomly selected from the replay memory, grouped into a batch, and fed into the reinforcement learning model for training. This process of random sampling effectively disrupts the time dependence between consecutive samples, thereby enhancing the efficiency of the training process. On the other hand, target network is used as a copy of the primary policy model for caching replacement, where its parameters are frozen for N steps and then updated to match the weights of the primary policy model, which can be viewed as a neural network to be optimized at each step. In particular, the target network provides stable target values during the training process, which helps to address the issue of catastrophic forgetting. This problem arises due to the frequent updates of the primary policy model, potentially destabilizing the action-value estimates for the subsequent state. The slowly-changed weights of the target network provide more consistent action-value estimates and enhance the stability and quality of the learned caching policies. Specifically, the target network is formulated by,

$$Y_i = r_i + \gamma \cdot \max_{a \in A} Q_{\theta^\star}(s_i', a), \tag{3.35}$$

where i is the client index and $\theta^\star$ is the current estimate of weight. The target network is then used to update the following gradient in every decision epoch,

$$L(\theta) = \frac{1}{n} \sum_{i=1}^{n} [Y_i - Q_\theta(s_i, a_i)]^2. \tag{3.36}$$

(2) DTN and safety module integration: First, DTN can be employed as both an RL optimizer and a safeguard, where real-time synchronization can be ensured through two-way communications between the physical network and DTNs as introduced in Sec. 3.2-B. For the sake of efficient computations, the current data is fed into the RL model to enhance training results that serve as the states in the

algorithm. To expedite the training process, logistic regression can be exploited to classify various states. Then, here the DTN is utilized for predictive purposes by simulating various twinning scenarios to anticipate and mitigate safety risks. These controllable scenarios help explore rare traffic cases and distributions that the network may encounter in the future, for example, DTN can generate a variety of distributions of client requests, enabling the RL agent to gain more knowledge and experience in handling diverse scenario cases.

Second, such an DTN-assisted approach can be further enhanced by incorporating the safety modules to proactively handle potential risks and ensure system reliability. In this section, we define the state s as $(n,\ t_n,\ f_n)$, which represents the BS index, the last time that the cache line was cached, and its caching frequency, respectively. Here, a safety module is added to the state space of our RL model. For instance, two additional variables can be incorporated into the cache state, namely, the client index k that is currently requesting content, and the normalized load d_n for the n-th BS. The normalized load d_n is the proportion of requests handled by the n-th BS out of all requests at time t. These additional variables force the policy model to consider the load of BSs. Therefore, the RL model can detect potential overloading issues of BSs, leading to more effective and balanced cache management. On the other hand, an additional penalty term can be integrated into the original reward function to further safeguard the caching process. In specific, after computing the reward in the DTN, the model can be penalized by the average difference between n-th BS and the minimum BS load, which can be formulated as follow:

$$r(s, a, s')' = r(s, a, s') - \frac{\phi}{N} \sum_n (d_n - \min_j d_j), \tag{3.37}$$

where N is the total count of BSs and ϕ is a hyperparameter to balance the significance of this additional term. Such a penalty term instructs the RL model to prioritize both balance in BS load and caching efficiency throughout the training process.

In summary, by incorporating safety modules and DTNs into the RL decision-making process, the wireless caching can be well operated while minimizing the likelihood of network failures. Notably, the inclusion of DTNs as reinforcement learning optimizers and protectors in wireless networks facilitates the use of diverse data patterns for the training and evaluation of cache replacement policies, which improves network adaptability and efficacy in managing cached content across densely deployed small-cell BSs.

3.5 Conclusions and Discussions

In this chapter, the creation and synchronization of DTNs are discussed with using ML approaches. The interplay between DTNs and ML includes two aspects, DTNs for ML and ML for DTNs. Considering DTNs over actual wireless communication scenarios, the joint communication and computation resource are optimized for improving the ML performance. Besides, the advantages of using RL for solving the problems in DTNs are provided in details. Future directions about ML assisted DTNs include the large ML model for DTNs over wireless communication, integrated sensing, communication, and computation for DTNs.

References

1. M. Raza, P.M. Kumar, D.V. Hung, W. Davis, H. Nguyen, R. Trestian, A digital twin framework for industry 4.0 enabling next-gen manufacturing, in *IEEE International Conference on Industrial Technology and Management (ICITM)* (2020)
2. C. Cimino, E. Negri, L. Fumagalli, Review of digital twin applications in manufacturing. Comput. Ind. **113**, 103130 (2019)
3. W. Kritzinger, M. Karner, G. Traar, J. Henjes, W. Sihn, Digital twin in manufacturing: a categorical literature review and classification. IFAC-PapersOnline **51**(11), 1016–1022 (2018)
4. M. Grieves, J. Vickers, Digital twin: mitigating unpredictable, undesirable emergent behavior in complex systems, in *Transdisciplinary Perspectives on Complex Systems* (Springer, Berlin, 2017), pp. 85–113
5. J. Yu, Y. Song, D. Tang, J. Dai, A digital twin approach based on nonparametric Bayesian network for complex system health monitoring. J. Manufact. Syst. **58**, 293–304 (2021)
6. L. Jakovljevic, D. Kostadinov, A. Aghasaryan, T. Palpanas, Towards building a digital twin of complex system using causal modelling, in *International Conference on Complex Networks and Their Applications* (Springer, Berlin, 2021), pp. 475–486
7. Siemens, Digital twin (2022). https://www.plm.automation.siemens.com/global/en/ourstory/glossary/digital-twin/24465
8. F. Tao, F. Sui, A. Liu, Q. Qi, M. Zhang, B. Song, Z. Guo, S.C.Y. Lu, A.Y. Nee, Digital twin-driven product design framework. Int. J. Product. Res. **57**(12), 3935–3953 (2019)
9. C. Lo, C. Chen, R.Y. Zhong, A review of digital twin in product design and development. Adv. Eng. Inf. **48**, 101297 (2021)
10. E. Negri, L. Fumagalli, M. Macchi, A review of the roles of digital twin in CPS-based production systems. Proc. Manufact. **11**, 939–948 (2017)
11. J. Vachálek, L. Bartalský, O. Rovný, D. Šišmišová, M. Morháč, M. Lokšík, The digital twin of an industrial production line within the industry 4.0 concept, in *2017 21st International Conference on Process Control (PC)* (IEEE, Piscataway, 2017)
12. F. Longo, L. Nicoletti, A. Padovano, Ubiquitous knowledge empowers the smart factory: the impacts of a service-oriented digital twin on enterprises' performance. Ann. Rev. Control **47**, 221–236 (2019)
13. M. Schluse, J. Rossmann, From simulation to experimentable digital twins: simulation-based development and operation of complex technical systems, in *2016 IEEE International Symposium on Systems Engineering (ISSE)* (IEEE, Piscataway, 2016), pp. 1–6
14. S. Boschert, R. Rosen, Digital twin—the simulation aspect, in *Mechatronic Futures* (Springer, Berlin, 2016), pp. 59–74.
15. H.X. Nguyen, R. Trestian, D. To, M. Tatipamula, Digital twin for 5G and beyond. IEEE Commun. Mag. **59**, 10–15 (2021)
16. Y. Wu, K. Zhang, Y. Zhang, Digital twin networks: a survey. IEEE Internet Things J. **8**(18), 13789–13804 (2021)
17. Y. Lu, X. Huang, K. Zhang, S. Maharjan, Y. Zhang, Communication-efficient federated learning for digital twin edge networks in industrial IoT. IEEE Trans. Ind. Inf. **17**(8), 5709–5718 (2020)

18. L.U. Khan, W. Saad, D. Niyato, Z. Han, C.S. Hong, Digital-twin-enabled 6G: vision, architectural trends, and future directions. IEEE Commun. Mag. **60**(1), 74–80 (2022)
19. S. Tao, Z. Cheng, D. Xiao-Dong, L. Lu, C. Dan-Yang, Y. Hong-Wei, Z. Yan-Hong, L. Chao, L. Qin, W. Xiao, et al., Digital twin network (DTN): concepts, architecture, and key technologies. Acta Autom. Sin. **47**(3), 569–582 (2021)
20. M. Mashaly, Connecting the twins: a review on digital twin technology & its networking requirements. Proc. Comput. Sci. **184**, 299–305 (2021)
21. A. Moser, C. Appl, S. Brüning, V.C. Hass, Mechanistic mathematical models as a basis for digital twins, in *Digital Twins: Tools and Concepts for Smart Biomanufacturing* (Springer, Berlin, 2021), pp. 133–180
22. M. Lohtander, N. Ahonen, M. Lanz, J. Ratava, J. Kaakkunen, Micro manufacturing unit and the corresponding 3d-model for the digital twin. Proc. Manufact. **25**, 55–61 (2018)
23. Z. Qing, Z. Liguo, D. Yulin, H. Han, G. Xuming, L. Mingwei, W. Wei, From real 3d modeling to digital twin modeling. Acta Geod. Cartograph. Sin. **51**(6), 1040 (2022)
24. E.N. Almeida, M. Rushad, S.R. Kota, A. Nambiar, H.L. Harti, C. Gupta, D. Waseem, G. Santos, H. Fontes, R. Campos, et al., Machine learning based propagation loss module for enabling digital twins of wireless networks in ns-3 (2022). arXiv:2205.04284
25. H. Elayan, M. Aloqaily, M. Guizani, Digital twin for intelligent context-aware IoT healthcare systems. IEEE Internet Things J. **8**(23), 16749–16757 (2021)
26. M.J. Kaur, V.P. Mishra, P. Maheshwari, The convergence of digital twin, IoT, and machine learning: transforming data into action, in *Digital twin technologies and smart cities* (Springer, Berlin, 2020), pp. 3–17
27. Y. Liu, D.M. Blough, Environment-aware link quality prediction for millimeter-wave wireless lans, in *Proceedings of the 20th ACM International Symposium on Mobility Management and Wireless Access* (2022), pp. 1–10
28. Y. Liu, S.K. Crisp, D.M. Blough, Performance study of statistical and deterministic channel models for mmWave wi-fi networks in ns-3, in *Proceedings of the Workshop on ns-3* (2021), pp. 33–40
29. H. Assasa, J. Widmer, T. Ropitault, N. Golmie, Enhancing the ns-3 IEEE 802.11 ad model fidelity: beam codebooks, multi-antenna beamforming training, and quasi-deterministic mmwave channel, in *Proceedings of Workshop on ns-3* (2019)
30. Y. Liu, Y. Jian, R. Sivakumar, D.M. Blough, Maximizing line-of-sight coverage for mmWave wireless LANs with multiple access points. IEEE/ACM Trans. Netw. **30**(2), 698–716 (2021)
31. Y. Liu, M. Chen, D. Xu, Z. Yang, S. Zhao, E-app: an environment-aware access point planning framework for mmwave wireless lans, in *Proceedings of IEEE International Conference on Computer Communications and Networks* (2023)
32. Y. Liu, Y. Jian, R. Sivakumar, D.M. Blough, Optimal access point placement for multi-AP mmWave WLANS, in *Proceedings of International ACM Conference on Modeling, Analysis and Simulation of Wireless and Mobile Systems* (2019)
33. J. Zhou, G. Cui, S. Hu, Z. Zhang, C. Yang, Z. Liu, L. Wang, C. Li, M. Sun, Graph neural networks: a review of methods and applications. AI Open **1**, 57–81 (2020)
34. D. Zhang, J. Yin, X. Zhu, C. Zhang, Network representation learning: a survey. IEEE Trans. Big Data **6**(1), 3–28 (2018)
35. D. Maltoni, V. Lomonaco, Continuous learning in single-incremental-task scenarios. Neural Netw. **116**, 56–73 (2019)
36. O. Hashash, C. Chaccour, W. Saad, K. Sakaguchi, T. Yu, Towards a decentralized metaverse: synchronized orchestration of digital twins and sub-metaverses (2022). arXiv:2211.14686
37. D. Balduzzi, Grammars for games: a gradient-based, game-theoretic framework for optimization in deep learning. Front. Robot. AI **2**, 39 (2016)
38. M. Chen, Z. Yang, W. Saad, C. Yin, H.V. Poor, S. Cui, A joint learning and communications framework for federated learning over wireless networks. IEEE Trans. Wireless Commun. **20**(1), 269–283 (2021)
39. O. Hashash, C. Chaccour, W. Saad, Edge continual learning for dynamic digital twins over wireless networks (2022). arXiv:2204.04795

40. S. Boyd, S.P. Boyd, L. Vandenberghe, *Convex Optimization.* Cambridge University Press, Cambridge (2004)
41. M.S. Ali, E. Hossain, A. Al-Dweik, D.I. Kim, Downlink power allocation for comp-noma in multi-cell networks. IEEE Trans. Commun. **66**(9), 3982–3998 (2018)
42. Y. Lu, S. Maharjan, Y. Zhang, Adaptive edge association for wireless digital twin networks in 6G. IEEE Internet Things J. **8**(22), 16219–16230 (2021)
43. M. Li, J. Gao, C. Zhou, X. Shen, W. Zhuang, User dynamics-aware edge caching and computing for mobile virtual reality. IEEE J. Sel. Top. Signal Process. **17**, 1131–1146 (2023)
44. C. Li, L. Toni, J. Zou, H. Xiong, P. Frossard, Qoe-driven mobile edge caching placement for adaptive video streaming. IEEE Trans. Multimedia **20**(4), 965–984 (2018)
45. L. Zhao, H. Li, N. Lin, M. Lin, C. Fan, J. Shi, Intelligent content caching strategy in autonomous driving toward 6G. IEEE Trans. Intell. Transp. Syst. **23**(7), 9786–9796 (2022)
46. V. Mnih, Playing atari with deep reinforcement learning (2013). arXiv:1312.5602
47. V. Mnih, K. Kavukcuoglu, D. Silver, A.A. Rusu, J. Veness, M.G. Bellemare, A. Graves, M. Riedmiller, A.K. Fidjeland, G. Ostrovski, S. Petersen, C. Beattie, A. Sadik, I. Antonoglou, H. King, D. Kumaran, D. Wierstra, S. Legg, D. Hassabis, Human-level control through deep reinforcement learning. Nature **518**(7540), 529–533 (2015). http://dx.doi.org/10.1038/nature14236

Chapter 4
Optimal Offloading in Digital Twin-Assisted Multi-Stage Networks

Muhammad Adnan Qadir, Muhammad Naeem, and Waleed Ejaz

4.1 Introduction

Due to the increase of computationally complex and time-constrained services in 6th generation (6G) devices, there is a requirement for multi-stage processing. Multi-stage processing is useful in dealing with time-constrained tasks by utilizing parallel processing. The computation of the same task at multiple stages simultaneously is called parallel processing. The ultra-reliable and low-latency communication (URLLC) protocol serves the purpose of low latency and high reliability in communication. By URLLC protocol in the multi-stage network, mission-critical applications with strict latency requirements and high block error rates are enabled [1].

The multi-stage network is utilized for parallel processing in multiple stages. Three layers are considered in the multi-stage network, with the addition of two extra layers of the edge and cloud server layers to the user layer. The UAVs are considered edge servers for mobile edge computing (MEC) due to their flexible mobility. UAVs are handy for their flexible placement, so they can be utilized for connectivity services in their coverage areas by acting as MEC servers [2]. MEC server is necessary for fast computing the users' time-sensitive and intensive computational tasks by providing the computation resources in the users' proximity. The UAV-MECs assisted communication enables the task computation with stringent latency and intensive computation resources requirement [3]. Cloud computing is necessary in multi-layer networks to facilitate highly computational

M. A. Qadir · W. Ejaz (✉)
Department of Electrical and Computer Engineering, Lakehead University, Thunder Bay Barrie/Orillia, ON, Canada
e-mail: mqadir1@lakeheadu.ca; wejaz@lakeheadu.ca

M. Naeem
Department of Electrical and Computer Engineering, COMSATS University Islamabad, Wah Campus, Punjab, Pakistan

M. K. Afzal et al. (eds.), *Digital Twins for Wireless Networks*,
https://doi.org/10.1007/978-3-031-73679-7_4

tasks. The cloud server is utilized to avoid the straggler effect from UAV-MECs [4]. The straggler effect can described as the users face delay if the task computation at UAV-MECs takes longer than the task arrival rate at UAV-MECs. The cloud server's computational capability can be utilized in this condition. In a multi-stage network, there should be an optimal division of tasks between the stages of processing for efficient utilization of resources.

Digital twin, introduced by Michael Grieves [5], is the virtual replica of physical objects to achieve the desired function from the physical objects through monitoring and control. A digital twin enables the virtual model to copy a physical world system accurately with the help of continuous iterative updates through interaction between the digital twin model and the physical model [6]. The digital twin technology enables the bidirectional information flow between the physical objects to the digital twin for updating and from the digital twin to the physical object for adaptive control [5].

The digital twin is a crucial tool to achieve control of physical objects through system modeling, real-time data processing, edge computing, and cloud computing [2]. A promising solution for efficient resource management and optimal network configuration in multi-stage networks can be achieved through digital twin [7]. The digital twin is an emerging technology enabling the virtual replica of physical objects for real-time management [1]. In multi-layer networks, it generates the virtual copy of wireless networks that can be utilized for the adaptive edge association for the users, enabling the path to sixth-generation (6G) networks [8].

4.1.1 Related Works

The URLLC is a necessary protocol for future-generation networks to ensure connectivity and low latency. In [9] the latency minimization problem is solved for an edge-cloud multi-layer network with URLLC. For the URLLC edge computing, the extreme value theory is applied in [10] by imposing the constraints for edge association to minimize the energy consumption by the users.

Multi-stage networks are utilized for parallel processing to reduce the processing latency for the user. The authors of [11] proposed a joint optimization in a three-layered architecture with optimal offloading and resource allocation to reduce energy consumption with the constraints of service latency. A survey is presented for different configurations of edge computation in [12]. In [12], authors also discussed resource scheduling in edge computation to highlight the usefulness of MECs. To minimize energy consumption by the dense multi-device system, a joint optimization problem is solved in [13] through optimizing the device association with the MEC server, task offloading, and resource allocation to devices and servers. In [9], the latency minimization problem is solved for a multi-stage network. The authors of [11] proposed a joint optimization in a multi-stage network with optimal offloading and resource allocation to reduce energy consumption. To minimize energy consumption by the dense multi-device system, a joint optimization problem is solved in [13] through optimizing the device association with the UAV-MECs, task offloading, and resource allocation to devices and servers. In [14], a multi-stage network is considered to optimize communication and computation resources. The authors of [1] considered the digital twin-assisted UAV-MECs network for latency minimization by optimizing the association, resource allocation, and task offloading.

A comprehensive survey on digital twin networks is presented in [5], highlighting the definition, enabling technologies, issues, and the usefulness of digital twins. Digital twins in wireless networks generate a virtual copy of wireless networks. The authors of [8] utilized the digital twin for the adaptive edge association for the users. The issue of movement and unpredictability of users in MEC networks is solved by utilizing the digital twin in [15] to reduce the offloading latency. Through digital twin in [7], the intelligent decision-making power of digital twin is utilized for the mobile edge selection in offloading the task to MECs. In [3], the digital twin is empowered with UAV-MECs and URLLC for enabling the 6G computation with low latency. The authors in [16] used the digital twin-assisted UAV-MECs networks to reduce the latency in an industrial IoT environment. For industrial automation, a digital twin-assisted UAV-MECs network is presented in [2] to minimize the end-to-end latency of IoT devices. Also, the same network utilized in [3] for enabling the 6G by minimizing the end-to-end latency for the users. Beyond the above background summarized in Table 4.1, this chapter proposes a digital twin-assisted multi-stage network with UAV-MECs and a cloud server to reduce mobile devices' latency to enable the environment of 6G.

The following are the main contributions of this chapter:

- We formulate the mathematical framework for optimizing task association and offloading portioning for the proposed multi-stage network through digital twin.
- A sequential solution is proposed to solve the optimization problem by dividing the problem into two sub-problems: association optimization through the K-means learning algorithm and offloading portioning optimization through the interior point algorithm.
- The simulation results for different scenarios are presented to emphasize the usefulness of the digital twin in multi-stage networks.

The rest of the chapter is organized as follows: Sect. 4.2 presents the system model and the optimization problem. The proposed solution is discussed in Sect. 4.3. Section 4.4 presents simulation results to show the effectiveness of the proposed scheme. The conclusion is drawn in Sect. 4.5.

4.2 System Model

We consider a digital twin-assisted offloading in a multi-stage network with U number of mobile users and N number of UAV-MECs for low-latency computation of the tasks by the users, as shown in Fig. 4.1. The considered environment is specifically designed for the limited users in remote areas. The set of U =

Table 4.1 Related Works

Reference	Publication	Digital twin	MECs/UAVs	Multi-stage network	Objective
[1]	2022	✓	MECs	Two-stages	Minimizing the worst case latency
[2]	2022	✓	UAVs	Two-stages	Minimizing the IoT devices task offloading latency
[3]	2022	✓	UAVs	Two-stages	Minimizing the end-to-end latency
[4]	2023	✗	MECs	Three-stages	Latency minimization of IoT devices task offloading and computation
[5]	2021	✓	✗	✗	A survey on digital twin network
[7]	2022	✓	MECs	Two-stages	Through intelligent offloading minimizing the latency
[8]	2021	✓	MECs	Two-stages	Adaptive edge association for mobile users
[9]	2022	✗	MECs	Three-stages	Minimizing offloading latency
[10]	2019	✗	MECs	Two-stages	Minimizing the power utilization by the users
[11]	2021	✗	MECs	Three-stages	Minimizing the energy consumption by the users
[12]	2021	✗	MECs	Two-stages	optimal resource scheduling in edge networks
[13]	2022	✗	MECs	Two-stages	Minimizing the energy consumption of system
[14]	2022	✓	MECs	Two-stages	Reducing end-to-end latency for metaverse environment
[15]	2020	✓	MECs	Two-stages	Reducing the offloading latency
[16]	2022	✓	MECs	Two-stages	Reducing industrial IoT devices offloading and computation latency
Proposed framework	TBD	✓	UAVs	Three-stages	Minimizing the digital-twin assisted three-stage network tasks' latency for users

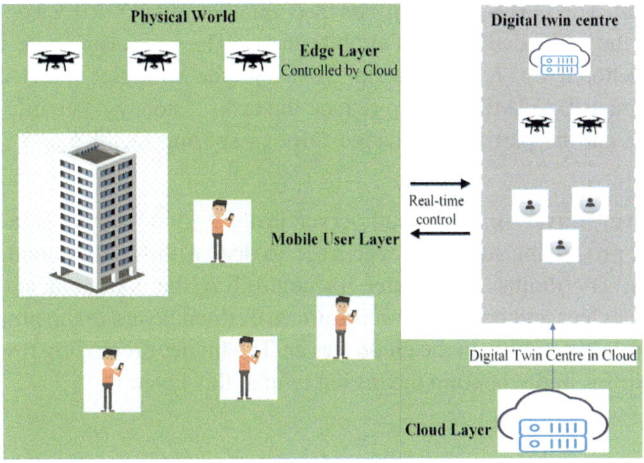

Fig. 4.1 System model for digital twin-assisted multi-stage network

$\{1, 2, \ldots U\}$ of U Mobile users (UEs) distributed in the environment and the set of $N = \{1, 2, \ldots N\}$ of N UAVs acting as mobile edge computing servers. A cloud server C with high computation resources is available to handle computationally complex tasks and acts as a digital twin control center as well. We assumed computation resources constrain the mobile devices to compute their task locally; then, the task will be offloaded to the other nearby layers with high computation power to achieve low computational latency. A task generated at u-th mobile user with the three components represented as $J_u = [D_u, C_u, T_u^{MAX}]$, with the task size of D_u, computational CPU cycles requirements of C_u, and the maximum allowable latency to complete the task T_u^{MAX}.

The task of mobile users can be partially computed with the local processing rate of f_u^L and shared with the UAV-MECs with the distribution factor of $\beta_u \in [0, 1]$. A binary variable $x_{u,n}$ for the association of u-th mobile user to the n-th UAV-MECs with the $x_{u,n} = 1$ the u-th mobile user is associated with the n-th UAV-MEC for offloading and $x_{u,n} = 0$, no association. It is assumed that $\sum_{n \in N} x_{un} = 1, \forall u$, i.e., the mobile user can offload a portion of its task to only one UAV-MEC at one time. A UAV-MEC can be associated with multiple mobile users who can be further connected to the cloud server. There is one more distribution factor $\alpha_{un} \in [0, 1]$ for deciding the offloading of tasks from n-th UAV-MEC to the cloud server.

The multi-stage network is classified into the following layers:

Mobile User Layer The mobile user layer consists of U number of mobile users. The task tuples are generated on this layer from mobile users, which can be locally computed with the processing rate f_u^L and offloaded as well to UAV-MECs with the distribution factor $\beta_u \in [0, 1]$. We assume that the tasks of mobile users are granular for offloading to achieve the low latency of communication and computation [1].

UAVs Layer The UAVs layer consists of N number of UAV-MECs responsible for processing the offloaded tasks of mobile users. The processing rate of nth UAV-MECs is denoted by f_n^E cycles/sec. To avoid the straggler effect due to tasks accumulation at UAV-MEC and to reduce the tasks' latency of mobile devices, the task portion can be partially offloaded through sharing factor $\alpha_{un} \in [0, 1]$ to the cloud server.

Cloud Server Layer In the cloud server layer, a cloud server is available with higher data processing rates f^C cycles/sec to deal with highly complex computational tasks. The digital twin center for controlling the communication resources for mobile devices and computation resources of three layers in the proposed multi-layer network is available in the cloud server. The higher processing resources make it the best fit for implementing the digital twin at this layer.

4.2.1 Communication Model

The channel model between the UAV-MECs and the mobile users has the components of line of sight (LOS) propagation, attenuation, blockage, and shadowing [2]. Here, we assumed that there would be a LOS link between the mobile user and UAV-MEC. So, the LOS model for the channel between the n-th UAV-MEC and u-th mobile user is given as [17]:

$$h_{un} = \beta_o d_{un}^{-\zeta} \tag{4.1}$$

where β_o denotes the reference channel gain at the reference distance of 1m, and $\zeta \geq 2$ denotes the path loss exponent [17]. d_{un} represents the Euclidean distance between the n-th UAV-MECs and u-th mobile user: $d_{un} = \sqrt{e_{un}^2 + l_k^2}$, where e_{un} denotes the horizontal distance between the u-th mobile user and n-th UAV-MECs, and l_n^2 represents height of the n-th UAV-MECs.

By utilizing the simplified expression of data rate between the u-th mobile user associated with n-th UAV-MEC can be approximated by Huynh et al. [1]:

$$R_u(p, x_u) = Blog_2[1 + \gamma_{un}(p, x_u)], \tag{4.2}$$

where B represents the bandwidth available for the transmission and $\gamma_{un}(p, x_u)$ represents the signal to noise ratio.

4.2.2 Digital Twin-Based Computation Model

The digital twin will generate the virtual copy of the whole multi-layer network, which will be represented as:

$$DT = \{\bar{U}, \bar{N}, \bar{C}\}, \tag{4.3}$$

The $\{\bar{U}, \bar{N}, \bar{C}\}$ are the virtual replicas of U IoT devices, N UAV-MECs, and C a cloud server, respectively. We can control and manage the physical system through the digital twin by exchanging the data between the model and the digital twin instance [7]. There are different tools available to implement the digital twin like the Modelica, DELMIA, FlexSim, etc. [1]. The digital twin model of the u-th mobile user is denoted by DT_u^L which is given as:

$$DT_u^L = (f_u^L, \hat{f}_u^L), \tag{4.4}$$

where f_u^L denotes the estimated processing rate of mobile users at digital twin and $\hat{f}_u^L$ represents the error in the approximation.

The digital twin virtual model for the n-th UAV-MECs is denoted by DT_n^E and given as:

$$DT_n^E = (f_n^E, \hat{f}_n^E), \tag{4.5}$$

where f_n^E represents the estimated processing rate and $\hat{f}_n^E$ as the approximation error. The digital twin virtual model for the cloud is denoted by DT^C, which is given as:

$$DT^C = (f^C, \hat{f}^C), \tag{4.6}$$

where f^C denotes the estimated processing rate, and $\hat{f}^C$ represents the approximation error. We now model the overall latency of the considered multi-layer network: local processing, task offloading to UAV-MECs, UAV-MECs processing, task Offloading to the cloud, and cloud processing. We assumed that the downlink transmission rate is higher and that downlink latency after processing can be ignored [1]. The data size after processing of the computation task is typically small and the UAV-MECs can transmit with high power; therefore, the downlink transmission latency can be ignored.

4.2.2.1 Local Processing Latency

The u-th mobile device is used to partially offload its task's portion β_u to the UAV-MECs as per the latency requirement. Assuming that the digital twin approximation error is known earlier, the error in latencies can be calculated. Thus, the actual latency of local computation will be the sum of the approximated latency and the approximation error [1]:

$$t_u^L(\beta_u) = \frac{(1-\beta_u)C_u}{f_u^L} + \Delta t_u^L(\beta_u). \tag{4.7}$$

4.2.2.2 Offloading Latency

After the association the task portion β_u from the u-th mobile user will be offloaded to UAV-MEC, so the offloading latency can be written as [1]:

$$t_u^O(\beta_u, x_u) = \frac{\beta_u D_u}{R_u(p, x_u)}. \tag{4.8}$$

4.2.2.3 UAV-MECs Processing Latency

Similar to the digital twin model of mobile users, the error in processing latency and its digital twin estimation can be calculated. So, the actual digital twin latency to compute the offloaded tasks at UAV-MECs can be expressed as [1]:

$$t^E = \frac{\sum_{u \in U} x_{un} \beta_u (1 - \alpha_{un}) C_u}{f_n^E} + \Delta t^E. \tag{4.9}$$

4.2.2.4 Offloading Latency UAV-MECs to Cloud

Each UAV-MECs offloads the tasks portion specified by $\alpha_{un} = [0, 1]$ to the cloud for a higher processing rate. The offloading latency between the UAV-MECs and the cloud can be written as [1]:

$$t^F(\beta, \alpha, x) = \sum_{u \in U} x_{un} \beta_u \alpha_{un} \frac{D_u}{R^F}, \tag{4.10}$$

where R^F represents the data rate of offloading the task from UAV-MECs to the cloud server.

4.2.2.5 Cloud Processing Latency

The error in the digital twin model's latency estimation can be calculated earlier. Thus, the actual latency of the cloud processing is given as:

$$t^C(\beta, \alpha, x) = \frac{\sum_{u \in U} \sum_{n \in N} x_{un} \beta_u \alpha_{un} C_u}{f^C} + \Delta t^C(\beta, \alpha, x). \tag{4.11}$$

From the above (4.7)–(4.11), total latency $t_u(\beta, \alpha, x)$ for the task of u-th mobile user will be the sum of computation, offloading, and synchronization latency.

The synchronization latency involved in sharing network configuration and other information between physical and digital twin models and digital twin model processing for intelligent decisions about offloading.

We mathematically formulate the optimization problem to minimize the latency of tasks from mobile users.

$$\min_{\beta,\alpha,x} : \sum_u t_u(\beta, \alpha, x)$$

$$\text{Subject to: C1–C7} \tag{4.12}$$

Input The input values which we give to this problem are the number of IoT devices U, the number of UAV-MECs N, task latency deadline T_u^{max}, and system bandwidth B.

Optimization Variables The variables that will be optimized in this problem are task portioning factors α_{un} & β_u and IoT devices association x_{un}.

Output In the optimization problem, we minimized the latency, jointly optimizing the network's computational and communication resources and association of IoT devices. The output will be the optimized variables mentioned above, resulting in minimized task latency.

Optimization Constraints

- **C1:** The total task computation and offloading latency $t_u(\beta, \alpha, x)$ should be less than the task's deadline T_u^{max}.
- **C2:** Quality of service ensures that the data rate of u-th IoT device in (4.2) should be greater than the minimum data rate represented by R_u^{min}.
- **C3:** Range of task sharing factor between u-th IoT device β_u to UAV-MECs should be between 0 and 1.
- **C4:** Range of task sharing factor of u-th IoT device between n-th UAV and cloud should be between 0 and 1.
- **C5:** Binary association variable can be 0 or 1.
- **C6:** The u-th IoT device can only be associated with one UAV-MEC at a given time.
- **C7:** The number of IoT devices that can be associated with the m-th UAV-MEC is bounded by the maximum association capacity of UAV-MECs represented by M^{max}.

4.3 Proposed Multi-Stage Solution

We initialized the parameters for the simulation in the MATLAB environment. The above optimization problem (4.12) is divided into two subproblems: the association optimization problem and the communication and computation resources optimization. To solve the first subproblem, we used the Kmeans placement algorithm for the UAVs in an environment depending upon the density of mobile users. The number of centroids found for mobile users is the same as the available UAVs. The centroid location is assigned to the UAV-MECs. Then, we made the association depending on the best channel conditions with constraints of $C5 - C7$.

$$\min_{\beta,\alpha} : \sum_{u} t_u(\beta, \alpha, x^*)$$
$$\text{Subject to: } C1 - C4. \tag{4.13}$$

After the association, the problem changes to (4.13) with the optimal association variable (x^*).

For the resources optimization, the interior point algorithm [18] is applied to the optimization problem (4.13). The solution region is defined for the interior point method, satisfying all the optimization constraints. Then the interior gradients are carried out until they change from negative to positive for all optimization variables to minimize the objective of task latency of the mobile users.

The complexity of the K-means algorithm is $O(UN)$, which is dependent on the number of mobile users (U) and number of UAV-MECs (N) [19]. The interior point algorithm is more complex with the complexity of $O(U^3N^3)$ [18]. The overall complexity for the above alternate optimization is $O(U^3N^3)$.

4.4 Simulation Results

Different simulations are carried out with varying numbers of mobile users and UAV-MECs to see the impact on the mobile users' task latency. The window of 100 $\times$ 100 m is considered, the mobile users are randomly distributed, and the UAV-MECs are placed by the K-means algorithm. The minimum data rate requirement of $R_u^{min} = 0.1$ Mbps is considered here with the C_u=800 Megacycles with maximum allowable latency $T_u^{max} = 2$ s. The digital twin error is considered as 3%, and the noise floor power is N_o=-174 dBm. We considered the simulation parameters similar to the [1].

Figure 4.2 shows the latency (seconds) versus the number of mobile users in the environment for different computational cycle requirements of tasks C_u. Here, the number of UAV-MECs are $N = 5$. The latency increases with the increment of the number of mobile users in the environment as the computation tasks increase with the number of users. The increment in computational cycles requirement C_u also increases the latency as it requires more time for the computation. It is very important to design this framework according to the users' density and the tasks computational cycles required to meet the stringent latency requirements of tasks.

Figure 4.3 shows the latency (seconds) plot versus the number of UAVs-MECs for the different computational cycles requirement of tasks by the users in the system. The latency decreases as the number of UAV-MECs equips the system with more computation resources to compute the tasks quickly. The latency reduces with the number of UAV-MECs, so it is very important to decide the number of UAV-MECs in the environment for the number of mobile users and the tasks computational cycles requirement C_u to meet the latency requirement.

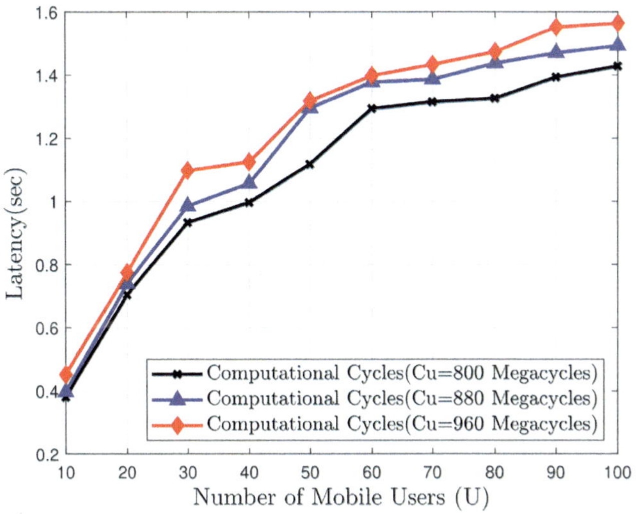

Fig. 4.2 Impact of number of mobile users on the tasks latency

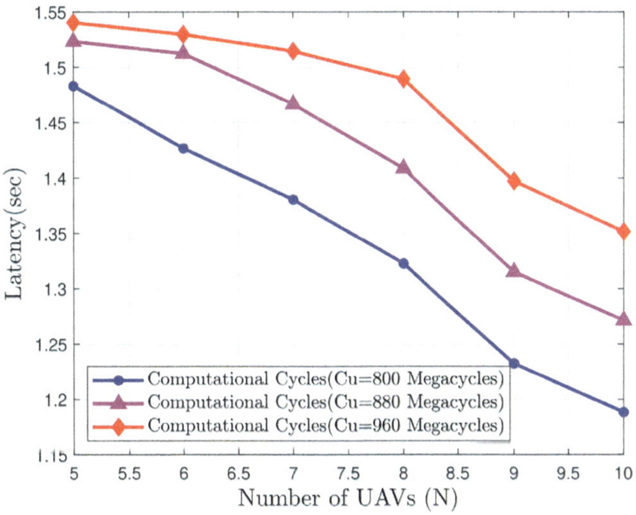

Fig. 4.3 Impact of number of mobile users on the tasks latency

4.5 Conclusion

In this book chapter, we formulated a latency minimization optimization problem in a digital-twin-assisted multi-stage network. Through utilizing the digital twin intelligence, we optimized the association of users with the UAV-MECs and offloading portioning of tasks between the different stages of the multi-stage network. The problem was solved through an alternate optimization; the Kmeans learning algorithm is utilized for association optimization, and the interior point method is used for offloading portioning optimization. We presented the simulation results to analyze the digital twin-assisted multi-stage network performance.

References

1. D.V. Huynh, V.D. Nguyen, S.R. Khosravirad, V. Sharma, O.A. Dobre, H. Shin, T.Q. Duong, URLLC edge networks with joint optimal user association, task Offloading and resource allocation: a digital twin approach. IEEE Trans. Commun. **70**(11), 7669–7682 (2022)
2. Y. Li, D.V. Huynh, T. Do-Duy, E. Garcia-Palacios, T.Q. Duong, Unmanned aerial vehicle aided edge networks with ultra-reliable low latency communications: a digital twin approach. IET Signal Process. **16**(8), 897–908 (2022)
3. T.Q. Duong, D.V. Huynh, Y. Li, E. Garcia-Palacios, K. Sun, Digital twin-enabled 6G aerial edge computing with ultra-reliable and low-latency communications, in *2022 1st International Conference on 6G Networking (6GNet)* (2022), pp. 1–5
4. D.V. Huynh, V.D. Nguyen, S. Chatzinotas, S.R. Khosravirad, H.V. Poor, T.Q. Duong, Joint communication and computation offloading for ultra-reliable and low-latency with multi-tier computing. IEEE J. Sel. Areas Commun. **41**(2), 521–537 (2023)
5. Y. Wu, K. Zhang, Y. Zhang, Digital twin networks: a survey. IEEE Internet Things J. **8**(18), 13789–13804 (2021)
6. L. Zhao, Z. Zhao, E. Zhang, A. Hawbani, A.Y. Al-Dubai, Z. Tan, A. Hussain, A digital twin-assisted intelligent partial offloading approach for vehicular edge computing. IEEE J. Sel. Areas Commun. **41**(11), 3386–3400 (2023)
7. T. Do-Duy, D.V. Huynh, O.A. Dobre, B. Canberk, T.Q. Duong, Digital twin-aided intelligent offloading with edge selection in mobile edge computing. IEEE Wireless Commun. Lett. **11**(4), 806–810 (2022)
8. Y. Lu, S. Maharjan, Y. Zhang, Adaptive edge association for wireless digital twin networks in 6G. IEEE Internet Things J. **8**(22), 16219–16230 (2021)
9. D.V. Huynh, V.D. Nguyen, S.R. Khosravirad, T.Q. Duong, Minimising offloading latency for edge-cloud systems with ultra-reliable and low-latency communications, in *ICC 2022 - IEEE International Conference on Communications* (2022), pp. 5122–5127
10. C.F. Liu, M. Bennis, M. Debbah, H.V. Poor, Dynamic task offloading and resource allocation for ultra-reliable low-latency edge computing. IEEE Trans. Commun. **67**(6), 4132–4150 (2019)
11. T.T. Vu, D.N. Nguyen, D.T. Hoang, E. Dutkiewicz, T.V. Nguyen, Optimal energy efficiency with delay constraints for multi-layer cooperative fog computing networks. IEEE Trans. Commun. **69**(6), 3911–3929 (2021)
12. Q. Luo, S. Hu, C. Li, G. Li, W. Shi, Resource scheduling in edge computing: a survey. IEEE Commun. Surv. Tuts. **23**(4), 2131–2165 (2021)
13. T. Zhou, Y. Yue, D. Qin, X. Nie, X. Li, C. Li, Joint device association, resource allocation, and computation offloading in ultradense multidevice and multitask IoT networks. IEEE Internet Things J. **9**(19), 18695–18709 (2022)

14. D.V. Huynh, S.R. Khosravirad, A. Masaracchia, O.A. Dobre, T.Q. Duong, Edge intelligence-based ultra-reliable and low-latency communications for digital twin-enabled metaverse. IEEE Wireless Commun. Lett. **11**(8), 1733–1737 (2022)
15. W. Sun, H. Zhang, R. Wang, Y. Zhang, Reducing offloading latency for digital twin edge networks in 6G. IEEE Trans. Veh. Technol. **69**(10), 12240–12251 (2020)
16. D.V. Huynh, V.D. Nguyen, V. Sharma, O.A. Dobre, T.Q. Duong, Digital twin empowered ultra-reliable and low-latency communications-based edge networks in industrial IoT environment, in *ICC 2022 - IEEE International Conference on Communications* (2022), pp. 5651–5656
17. J. Cui, Y. Liu, A. Nallanathan, Multi-agent reinforcement learning-based resource allocation for UAV networks. IEEE Trans. Wireless Commun. **19**(2), 729–743 (2020)
18. I. Pólik, T. Terlaky, *Interior Point Methods for Nonlinear Optimization* (Springer, Berlin, 2010), pp. 215–276
19. M.K. Pakhira, A linear time-complexity k-means algorithm using cluster shifting, in *2014 International Conference on Computational Intelligence and Communication Networks* (2014), pp. 1047–1051

Chapter 5
Employing Federated Learning for the Implication of Digital Twin

Fakhreldin Saeed, Momina Shaheen, Tariq Umer, and Muhammad S. Farooq

5.1 Introduction

The inception of the digital twin paradigm can be traced back to its roots in NASA's early space missions, where it was initially utilized for simulations and monitoring [1]. Today, digital twins have transcended their origins, finding a broad spectrum of applications spanning manufacturing [2, 3], healthcare [4, 5], smart cities [6], and beyond [7]. They are more than mere simulations; they are dynamic entities capable of continuously evolving alongside their real-world counterparts. At its core, a digital twin comprises three pivotal components: *a digital model*, *real-time data feeds*, and *advanced analytics*. The digital model serves as the backbone, encapsulating every detail of the physical asset or system. Real-time data streams, harvested from sensors, IoT devices, or other data sources, fuel the digital twin with a constant influx of information, allowing it to mirror real-world conditions [8]. The analytics component, powered by machine learning and artificial intelligence, equips the digital twin with the capability to make predictive decisions, optimize performance, and simulate scenarios for better decision-making [9].

The adoption of digital twins is widespread. In manufacturing, they are used to simulate and optimize production processes [3], predict equipment failures [10], and streamline operations [11]. In healthcare, digital twins of patients can help personalize treatment plans and predict disease outcomes [12]. Smart cities leverage

F. Saeed · M. Shaheen (✉) · T. Umer · M. S. Farooq
Computing and Digital Design, University of Roehampton, London, UK

Department of Computer Science, COMSATS University Islamabad, Islamabad, Pakistan

Department of Artificial Intelligence, School of Systems and Technology SST, University of Management and Technology, Lahore, Pakistan
e-mail: fakhreldin.saeed@roehampton.ac.uk; momina.shaheen@roehampton.ac.uk; tariqumer@cuilahore.edu.pk; shoaibfarooq@umt.edu.pk

M. K. Afzal et al. (eds.), *Digital Twins for Wireless Networks*,
https://doi.org/10.1007/978-3-031-73679-7_5

digital twins to optimize energy consumption, urban planning, and infrastructure maintenance. Transportation systems are advancing through the use of digital twins to enhance traffic management and logistics [13].

While digital twins empower the virtual representation and optimization of physical systems, federated learning introduces a powerful approach to machine learning [14]. It addresses the challenges posed by centralized models, where data from diverse sources is pooled into a single repository. Such a centralization can lead to data privacy concerns, scalability limitations, and significant communication overhead [15]. Federated learning, as a decentralized machine learning paradigm, diverges from this approach. It enables multiple edge devices or data sources to collaboratively train a global model without the necessity of sharing raw data. Local models, trained on the edge, participate in a collaborative learning process [16]. They compute local updates, which are shared and aggregated to form a global model. This mechanism preserves data privacy and reduces the need for extensive data transmission. The advantages of federated learning are compelling. It addresses data privacy concerns by ensuring that data never leaves its local source. Real-time insights can be derived by harnessing local knowledge. The system scales gracefully as new data sources are added, and it significantly reduces the communication overhead associated with centralized models. The applications of federated learning are extensive. Healthcare providers can collaboratively train models for disease diagnosis while protecting patient privacy. Autonomous vehicles can enhance their driving capabilities through collective learning from edge devices. Financial institutions can detect fraud and anomalies while keeping sensitive customer data secure. In the age of digital transformation, the convergence of two groundbreaking technologies, digital twins and federated learning, has unveiled a new horizon of possibilities. The fusion of digital twin technology with federated learning represents a synergy that is redefining the landscape of diverse industries [17–20]. This amalgamation holds immense promise in addressing critical challenges, particularly in scenarios where data privacy, decentralized data sources, real-time insights, and scalability are paramount concerns.

This chapter is devoted to exploring the implications of employing federated learning within the context of digital twins. We will delve into the rationale behind integrating federated learning and digital twins, showcase real-world applications and case studies, discuss the fundamental principles of data partitioning and model aggregation, and address privacy considerations. Additionally, we will examine the challenges faced and future directions for this transformative combination. With this chapter as a guide, readers will gain a comprehensive understanding of the potential that lies within the integration of federated learning in digital twin applications, serving as a springboard for further exploration and innovation.

5.2 Digital Twins: Foundations and Applications

5.2.1 Core Components of Digital Twins

To comprehend the essence of digital twins, it's essential to dissect their core components as represented in Fig. 5.1.

5.2.1.1 Digital Model

At the heart of every digital twin lies a meticulously crafted digital model [21]. This digital representation serves as a mirror image of the physical asset or system [22]. It encapsulates every intricate detail, from the geometry and structure to the operational characteristics with real-time data feed [23].

5.2.1.2 Real-Time Data Feeds

Digital twins thrive on real-time data [24]. This information is harvested from a myriad of sources, ranging from sensors, IoT devices, cameras, and other data-capturing mechanisms. These data feeds supply the digital twin with a continuous influx of information, allowing it to mirror real-world conditions with precision and accuracy.

5.2.1.3 Advanced Analytics

Digital twins are not static replicas but dynamic entities capable of analysing, simulating, and predicting real-world behaviour. This capability is made possible through advanced analytics, driven by machine learning, artificial intelligence, and

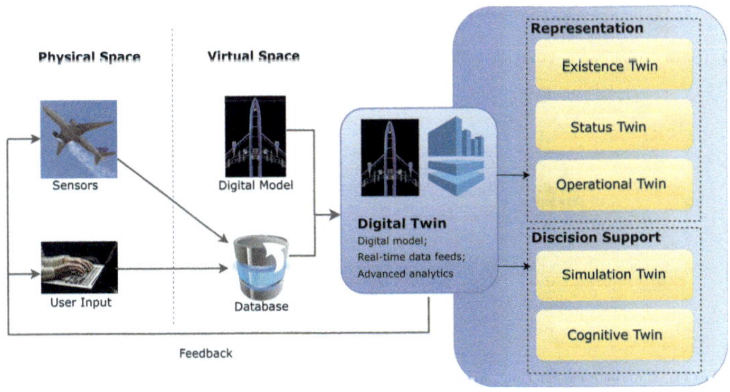

Fig. 5.1 Components of digital twin

data analytics techniques [25]. These technologies empower the digital twin to make predictive decisions, optimize performance, and simulate various scenarios.

5.2.2 Applications of Digital Twins

The versatility of digital twins is showcased through their applications in diverse sectors Fig. 5.2

5.2.2.1 Manufacturing

In the realm of manufacturing, digital twins are harnessed to optimize production processes. They simulate manufacturing operations, predict equipment failures, and provide real-time insights into operational efficiency [26]. Manufacturers use digital twins to refine product design, streamline production, and enhance quality control.

5.2.2.2 Healthcare

Healthcare practitioners leverage digital twins to create patient-specific models [27–29]. These models provide insights into disease progression, enabling personalized

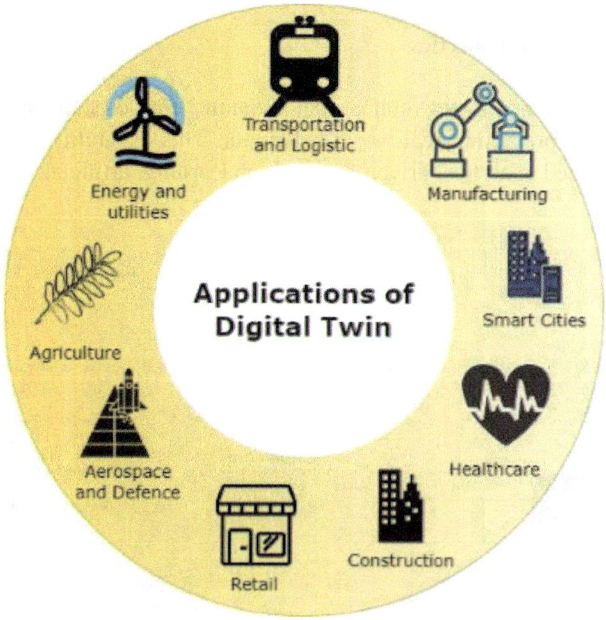

Fig. 5.2 Applications of digital twin

treatment plans [30, 31]. In the context of medical devices, digital twins are used for remote monitoring, ensuring the optimal performance of devices and early detection of issues [32].

5.2.2.3 Smart Cities

Smart cities rely on digital twins to optimize energy consumption, urban planning, and infrastructure maintenance [33, 34]. Digital twins of city systems, such as electrical grids and transportation networks, help manage resources efficiently, predict and mitigate infrastructure failures, and enhance overall urban sustainability.

5.2.2.4 Transportation and Logistics

In the transportation sector, digital twins enable the creation of highly accurate simulations of transportation systems [35]. This includes traffic management, logistics optimization, and predictive maintenance of vehicles and infrastructure [36, 37]. Similarly, autonomous vehicles also use digital twins to simulate driving scenarios and enhance decision-making [38].

5.2.2.5 Aerospace and Defence

Aerospace and defence industries employ digital twins to create virtual prototypes of aircraft, spacecraft, and defence systems [39–42]. This aids in design validation, performance optimization, and predictive maintenance, reducing costs and enhancing safety.

5.2.2.6 Energy and Utilities

The energy sector uses digital twins for real-time monitoring and predictive maintenance of power generation and distribution systems [43, 44]. These digital twins help utility companies optimize their operations, reduce downtime, and enhance grid resilience [45–47].

5.3 Federated Learning: A Decentralized Machine Learning Paradigm

5.3.1 Decentralization: The Essence of Federated Learning

Federated learning, at its core, represents a paradigm shift in the world of machine learning. Unlike traditional centralized machine learning approaches where data is

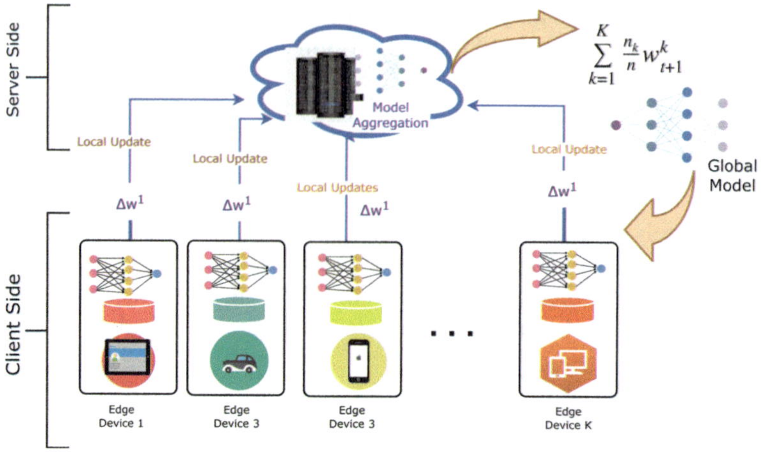

Fig. 5.3 Federated learning mechanism, where multiple edge nodes train a local model over the client side. The server is aggregating the model

collected and stored in a central repository for model training, federated learning is characterized by decentralization [16]. The fundamental premise is simple yet profound: the data stays where it originates, and the model is trained collaboratively across decentralized devices [48], as represented in Fig. 5.3. In a federated learning ecosystem, each edge device, such as a smartphone, IoT sensor, or local server, possesses its own local dataset. Instead of sharing this data with a central server, these local devices compute local model updates based on their data [49]. These updates are then sent to a central server or aggregator, where they are combined to form a global model. This global model is refined iteratively by aggregating and incorporating insights from all participating devices [50].

5.3.2 The Advantages of Federated Learning

Federated learning offers a range of compelling advantages that stem from its decentralized nature:

Data Privacy One of the most prominent advantages is the preservation of data privacy [51]. In a federated learning setup, raw data never leaves its local source. This is of paramount importance in applications where sensitive or confidential data is involved [52], such as healthcare, finance, and personal devices. Users can contribute to model training without compromising the security of their personal information [53].

Reduced Communication Overhead In contrast to centralized models that require the transmission of vast amounts of data to a central server, federated learning reduces communication overhead significantly. Only model updates, typically compact and encrypted, are exchanged between local devices and the central server [54]. This efficient communication minimizes bandwidth usage and latency, making it particularly suitable for real-time applications and low-bandwidth environments [55].

Scalability Federated learning systems are inherently scalable. As new devices or data sources are added to the network, they can seamlessly participate in the collaborative model training process [56, 57]. This ability to expand the network ensures that federated learning systems can adapt to changing data landscapes and remain versatile.

5.3.3 The Collaborative Learning Process

Federated learning, although decentralized, is a collaborative learning process. Local devices compute local model updates based on their data, leveraging the insights unique to their respective environments. These updates are then shared with a central server, where they are aggregated to create a global model [14]. This global model encapsulates the collective intelligence from all participating devices [58]. This collaborative approach allows federated learning to harness the collective knowledge from diverse sources without centralizing data. As we delve deeper into this chapter, we will explore how federated learning complements digital twins, providing solutions to issues such as data privacy, decentralized data sources, real-time insights, and scalability. The fusion of digital twins and federated learning is poised to redefine industries and empower organizations with a transformative approach to handling and harnessing data.

5.4 Why Federated Learning for Digital Twins?

As we navigate the landscape of digital twins and federated learning, it's crucial to understand the rationale behind their integration. Why do we turn to federated learning to enhance digital twin environments? The answer lies in the unique set of challenges faced by digital twin applications and how federated learning offers solutions to these challenges (as depicted in Fig. 5.4).

5.4.1 Data Privacy Concerns

Digital twins often deal with sensitive, proprietary, or personal data [59–61]. Whether it's manufacturing data, patient health records, or urban infrastructure details, maintaining data privacy is of paramount importance [60–63]. Centralized

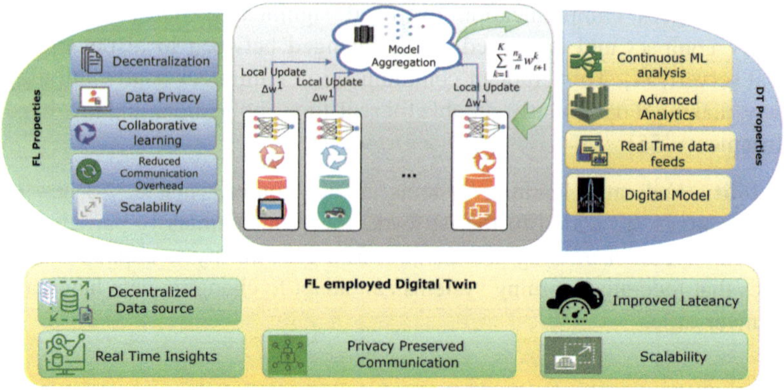

Fig. 5.4 Overview of FL based DT properties

machine learning approaches, where data is aggregated in a central repository, present a significant risk to data security and privacy. A breach of this central storehouse could lead to severe consequences. Federated learning addresses this concern by ensuring that data remains on the edge devices where it originates. Local models are trained with local data, and only the model updates, rather than raw data, are shared with a central server [49]. This approach ensures that sensitive information is never exposed, offering a robust solution to data privacy concerns in digital twin applications.

5.4.2 Decentralized Data Sources

Digital twin applications frequently draw data from decentralized sources. Edge devices, sensors, IoT endpoints, and distributed networks contribute data in real-time [59]. Traditional centralized models struggle to accommodate such diverse and distributed data sources efficiently. Federated learning thrives in this environment. It is designed to work with data sources that are distributed and heterogeneous [64]. Local devices are seen as equal contributors, each offering their unique insights derived from their specific environments. Federated learning enables the collaborative training of models, making it an ideal fit for digital twin applications that rely on data from decentralized sources [50].

5.4.3 Real-Time Insights

In many digital twin use cases, real-time insights are critical. Digital twins need to react and adapt to rapidly changing conditions in the physical world [8]. Centralized model training typically involves data transfer, model updates, and aggregation, which may introduce delays. These delays can be detrimental in scenarios where decisions must be made in real-time. Federated learning facilitates real-time insights [65]. Since local devices process data locally and share model updates, they can respond to changes swiftly. This real-time nature of federated learning aligns seamlessly with the requirements of digital twin applications [66], ensuring that insights are both timely and relevant.

5.4.4 Scalability

The growth and expansion of digital twin applications are often limited by the constraints of centralized systems [67]. Scaling up centralized models to accommodate additional data sources or devices can become unwieldy and resource-intensive.

Federated learning, on the other hand, is inherently scalable [68]. New edge devices can join the federated learning network with ease. They contribute their local knowledge to the collective model without requiring a complete overhaul of the system. This scalability ensures that digital twin applications can evolve and expand without encountering the challenges of centralized systems.

5.5 Related Work

To provide context for the integration of federated learning and digital twins, it is essential to review previous research and case studies that have demonstrated the feasibility and success of this combination. Key findings from related work in various industries, such as manufacturing, healthcare and smart cities, provide insights into the practical applications and achievements of this synergistic approach.

6G networks and federated learning can be utilized in the manufacturing industry to enhance digital twins. Federated learning enables the efficient updating of fault detection models on each server using distributed learning methods, which addresses the communication cost limitations of the centralized approach [50]. By utilizing federated learning, the progress of learning can be consolidated while preserving data ownership. This enables the incorporation of data from various sources without compromising intellectual property or legal constraints [69]. This approach is particularly beneficial in the industrial production environment, where data availability is limited due to various factors. The combination of federated learning and 6G network technology can enable seamless monitoring, control, and optimization of the manufacturing process through digital twins. This can lead to improved efficiency and productivity in the industry; on the other hand, the deployment of digital twins is enhancing the performance of the 6G infrastructure itself [70].

Digital twins in healthcare can benefit from the integration of 6G networks and federated learning. 6G networks, with their high-speed and low-latency capabilities, can enable real-time data transmission and communication between digital twins and healthcare systems, enhancing the accuracy and efficiency of healthcare applications. Federated learning, on the other hand, allows for collaborative model training without compromising data privacy, making it suitable for healthcare environments where data protection is crucial. By combining these technologies, digital twins can leverage the vast amount of data available in healthcare settings while ensuring privacy and security. This integration can enable personalized and precise healthcare interventions, such as patient-specific diagnosis and treatment planning, optimization of resource allocation and workflow processes in hospitals, and early detection and prediction of disease outbreaks [49, 71–74].

Smart cities, as urban centres that are increasingly employing digitization and innovative technologies, reap notable advantages from the convergence of 6G networks and federated learning. The domains of traffic management and environmental monitoring are prime beneficiaries of this fusion. The reduction of traffic congestion and the expedited commute times are tangible outcomes that enhance the urban experience. Federated learning contributes significantly to environmental monitoring, facilitating the mitigation of air pollution levels. This juncture underscores the transformative potential of digital twins, reinforced by federated learning, in the pursuit of sustainable and efficient urban ecosystems. In particular, the application of digital twins has led to improved traffic scene recognition. This is exemplified by the reduction in recognition times of prominent algorithms and the optimization of electricity usage alongside traffic flow forecasting. This collective progression signifies the promise inherent in the integration of digital twins and federated learning for building smart and eco-conscious urban landscapes [75, 76].

5.6 Implementation Guidelines

5.6.1 Data Partitioning in Digital Twins for Federated Learning

Effective data partitioning is the linchpin of successful federated learning in the context of digital twins [77]. In this section, we delve into the intricacies of data partitioning, understanding its importance and exploring strategies that enhance the efficiency and effectiveness of this process. Data partitioning ensures that digital twins can make the most of federated learning while addressing specific challenges unique to their applications [78].

5.6.1.1 The Importance of Data Partitioning

Data partitioning is the process of dividing the dataset among participating edge devices in a federated learning setup. The significance of this process cannot be overstated. Efficient data partitioning ensures that local devices have access to data that is relevant to their specific contexts and use cases. In the realm of digital twins, this ensures that each digital twin receives data that accurately reflects the conditions and characteristics of the physical asset it represents. Effective data partitioning also plays a vital role in optimizing the federated learning process [78]. It reduces communication overhead, minimizes the data transferred between devices and the central server, and allows for focused model updates that reflect localized insights. Ultimately, data partitioning is a key enabler of real-time, context-aware insights within digital twin applications.

5.6.1.2 Strategies for Effective Data Partitioning

1. **Data Volume-Based Partitioning**: Data volume is a critical consideration when partitioning data for federated learning in digital twins. It's essential to ensure that each edge device receives a sufficient amount of data to train a meaningful local model. Considerations should be made to balance the data volumes across devices to prevent overloading or underutilizing any one device.
2. **Data Quality-Driven Partitioning**: Data quality is another key dimension to consider when partitioning data. Digital twin applications rely on high-quality data to ensure accurate representations and insights. Prioritizing data quality in partitioning ensures that each device receives a dataset that meets the necessary quality standards, enabling the creation of robust digital twin models.
3. **Localization-Based Partitioning**: One of the most crucial aspects of data partitioning in digital twins is localization. Digital twins represent physical entities, and data that is localized to the geographical or spatial context of the asset is invaluable. Localization-based partitioning ensures that each digital twin receives data that accurately reflects its immediate environment. This enables the creation of digital twin models that are finely tuned to the specific conditions and characteristics of the assets they represent.

5.6.1.3 Challenges and Considerations

Data partitioning, while critical, is not without its challenges. In some scenarios, data may be unevenly distributed across edge devices, leading to imbalanced model training [16, 79]. Strategies for dealing with imbalanced data partitions should be developed. Additionally, effective data partitioning can be resource-intensive, especially when dealing with large and complex datasets. The design of data partitioning algorithms and strategies should consider the specifics of the digital twin application. Factors such as the spatial layout of the assets, the granularity of data, and the dynamics of the physical systems play a role in determining the most suitable partitioning approach.

5.6.2 Model Aggregation Techniques Used in Federated Learning

5.6.2.1 Federated Averaging

Federated Averaging is one of the fundamental model aggregation techniques in federated learning [80]. It operates on the principle of decentralized model updates from edge devices. Here's how it works:

- Local Model Updates: Each participating edge device computes its local model update based on its dataset [14]. This local model update captures the insights and knowledge from the specific data source.
- Model Averaging: The local model updates are then shared with a central server or aggregator, where they are aggregated to create a global model [14].

$$\omega = (1/N) \sum_{i=1}^{N} \omega_i$$

where

- ω: is the aggregate update
- N: is a set of participating clients
- ω_i: is client update

The most common approach is to perform a simple weighted average, where the contribution of each device's update is weighted by its data size or computational resources.

- Consensus Building: The global model is distributed back to the participating devices, aligning the local models with the global consensus. This process is typically repeated iteratively to refine the global model over multiple rounds of communication [15].

Algorithm 1 FedAvg (federated averaging)

Require: $K, T, \eta, E, w^0, N, p_k, k = 1, \ldots, N$
for t=0,$\cdots$,T-1 **do**
  Server selects a subset St of K devices at random(each device k is chosen with probability p_k)
  Server sends w^t to all chosen devices
  Each device $k \in S_t$ updates w^t for E epochs of SGD on F_k with step-size η to obtain w_k^{t+1}
  Each device $k \in S_t$ sends w_k^{t+1} back to the server
  Server aggregates the w's as $w^{t+1} = \frac{1}{k} \sum_{k \in S_t} w_k^{t+1}$
end for

5.6.2.2 Secure Aggregation

Secure Aggregation techniques focus on preserving data privacy during the model aggregation process [81]. They ensure that individual edge devices' contributions remain confidential. Secure Multi-Party Computation (SMPC) [82] and Homomorphic Encryption [83] are two common methods used for secure aggregation:

- Secure Multi-Party Computation (SMPC): SMPC enables multiple parties, in this case, the edge devices and the central server, to perform computations on their local models without revealing their raw data [84]. They collaborate to calculate the aggregated model while keeping their inputs private.
- Homomorphic Encryption: Homomorphic encryption allows computations to be performed on encrypted data [85]. Edge devices encrypt their local model updates before sharing them with the central server. The server can then perform operations on the encrypted data, leading to an aggregated model that is decrypted and shared with the devices.

Secure Aggregation techniques are crucial for applications where data privacy is a primary concern, such as healthcare and finance [81].

5.6.2.3 Model Update Compression

Model Update Compression techniques focus on reducing the communication overhead during model aggregation [86]. They are particularly valuable when dealing with limited bandwidth or high-latency communication links. The common method is Quantization. In quantization, the local model updates are reduced in precision [87]. For example, instead of transmitting model parameters as 32-bit floating-point numbers, they can be quantized to 8-bit integers. This significantly reduces the amount of data transmitted during aggregation. Model Update Compression techniques strike a balance between model accuracy and communication efficiency. They are crucial in scenarios where bandwidth and latency are constraints.

5.6.2.4 Differential Privacy

Differential Privacy is a technique that adds controlled noise to the local model updates before aggregation [88]. This noise ensures that the model updates do not reveal sensitive information about individual devices' data. While it increases privacy, it can also introduce trade-offs in model accuracy.

$$\omega = (1/N) \sum_{i=1}^{N} (\omega_i + b \cdot n_i)$$

where

- ω: is the aggregate update
- N: is a set of participating clients
- ω_i: is client update
- n_i: is a random noise vector, drawn from a Laplace distribution with a scale parameter "b".
- b: is a privacy budget parameter

5.6.2.5 Personalization and Adaptive Aggregation

Personalization and Adaptive Aggregation techniques focus on customizing the aggregation process based on the performance of individual edge devices. They adapt the aggregation method to the specific strengths and weaknesses of each device:

- Personalization: In personalization, the central server can assign different weights to each device's local model update based on their historical performance [58]. Devices that consistently provide high-quality updates receive more weight, while those with lower-quality updates receive less.
- Adaptive Aggregation: Adaptive aggregation techniques dynamically adjust the aggregation method based on real-time feedback [79]. For example, if a device's updates consistently degrade the global model's accuracy, the central server may reduce its influence in the aggregation process.

Personalization and Adaptive Aggregation techniques are valuable when dealing with heterogeneity in edge devices, ensuring that the global model benefits from the strengths of each device.

5.6.3 Ensuring Privacy in Federated Learning for Digital Twins

In the realm of federated learning, especially when integrated with digital twins, privacy takes centre stage. Preserving the confidentiality of sensitive data is paramount, and this section explores the critical role of privacy and the techniques used to safeguard it, with a particular focus on differential privacy [48].

5.6.3.1 The Crucial Role of Privacy

Digital twins often operate in contexts where data confidentiality is non-negotiable. Consider healthcare, where patient data is involved, or manufacturing, where proprietary processes are monitored. In these scenarios, maintaining privacy is critical not only for ethical reasons but also to comply with legal and regulatory requirements. Federated learning was designed, in part, to address these privacy concerns. Allowing data to remain on the edge devices and aggregating only model updates, significantly reduces the risk of data breaches [88]. However, to fortify privacy protections, advanced techniques are employed, with differential privacy at the forefront.

5.6.3.2 Differential Privacy (Protecting Individual Data)

In the context of federated learning for digital twins, it functions as follows:

- **Local Noise Addition:** Edge devices add random noise to their local model updates before sharing them with the central server. This noise masks any specific information about individual data points, ensuring that they remain confidential [89].
- **Aggregated Noise:** The central server, when aggregating the local updates, continues to add noise to the process. This aggregated noise provides a second layer of privacy protection, making it extremely challenging to reverse-engineer information about individual data sources [89].

Differential privacy is a robust approach to protect individual privacy [48]. It ensures that the insights gained from federated learning models do not come at the expense of data confidentiality. Its application in digital twin scenarios, where data can be highly sensitive, is invaluable.

5.7 Practical Implementation

5.7.1 Case Study: Implementing Federated Learning in a Smart Manufacturing Digital Twin

Smart manufacturing relies on the seamless integration of digital twins and real-world production processes. The fusion of these technologies offers the potential for optimization, predictive maintenance, and real-time insights [90, 91]. In this case study, we explore the practical implementation of federated learning within a smart manufacturing digital twin, detailing the data sources, model architecture, challenges faced, and the outcomes achieved.

5.7.1.1 Data Sources

- **Sensors and IoT Devices:** The manufacturing facility is equipped with a vast array of sensors and IoT devices [92] that collect data on equipment performance,

temperature, pressure, humidity, and other key metrics. These devices generate real-time data streams that serve as the primary data source.
- **Historical Production Data:** Historical production data, including equipment logs, maintenance records, and quality control data, provides essential context for understanding past performance and identifying potential improvements.
- **Supply Chain and Inventory Data:** Data related to supply chain logistics, raw material inventory, and product distribution are integrated to optimize production scheduling and resource allocation.

5.7.1.2 Model Architecture

The digital twin in this smart manufacturing environment [2] consists of the following key components:

- **Digital Model:** A detailed digital model is created to represent the manufacturing process. It includes a virtual representation of the equipment, production lines, and the product being manufactured. This model is constantly updated to mirror real-world conditions.
- **Federated Learning Framework:** The federated learning framework is deployed to enable model training across edge devices [65]. Edge devices in this case include the various machines and sensors on the production floor.
- **Local Models:** Each edge device, equipped with local data, maintains its own local model. These local models capture the specific performance characteristics of the respective device.
- **Aggregator and Global Model:** A central server acts as an aggregator. It receives local model updates, aggregates them, and computes a global model. The global model represents the collective insights from all edge devices.

5.7.1.3 Challenges Encountered

Implementing federated learning in a smart manufacturing digital twin environment presented several challenges:

- **Data Heterogeneity:** The data generated by different sensors and devices exhibited variations in format and granularity, requiring extensive data preprocessing efforts to ensure compatibility [16].
- **Real-Time Constraints:** The need for real-time insights demanded low-latency model updates. Balancing real-time processing with data privacy was a complex challenge.
- **Imbalanced Data Distribution:** Some machines had significantly more data points than others, leading to imbalanced contributions to the global model [93]. Strategies for dealing with this imbalance had to be developed.

5.7.1.4 Successful Outcomes

Despite the challenges, the implementation of federated learning in the smart manufacturing digital twin yielded significant benefits:

- **Real-Time Anomaly Detection:** The digital twin, powered by the global model, enabled real-time anomaly detection and predictive maintenance. Equipment failures were predicted before they occurred, reducing downtime and maintenance costs.
- **Production Optimization:** The global model provided insights into production efficiency and quality control. Adjustments to the manufacturing process were made in real-time to optimize production and reduce waste.
- **Data Privacy:** The privacy of sensitive manufacturing data was preserved, as data never left the edge devices. Differential privacy and secure aggregation techniques were successfully employed to protect data confidentiality [94].
- **Scalability:** As new machines and sensors were added to the production floor, the federated learning system seamlessly scaled to incorporate them into the model training process.

5.7.2 Case Study: Implementing Federated Learning in a Smart Cities Digital Twin

Smart Cities leverage advanced technologies to enhance urban living, optimize resource usage, and improve overall quality of life. In this case study, we explore the practical implementation of federated learning within a Smart Cities digital twin, detailing the data sources, model architecture, challenges encountered, and the outcomes achieved.

5.7.2.1 Data Sources

- IoT Devices and Sensors: Smart Cities are equipped with a network of IoT devices and sensors, including those for traffic management, environmental monitoring, waste management, and public infrastructure [78]. These devices generate real-time data, contributing to the city's digital twin.
- Urban Infrastructure Data: Information related to city infrastructure, including road networks, public transportation systems, and energy grids, is integrated. This data helps simulate and optimize the city's various systems.
- Citizen Feedback and Mobile Apps: Citizen feedback, collected through mobile apps and other platforms, provides valuable insights into urban experiences and issues. This data helps in identifying areas for improvement and enhancing city services.

5.7.2.2 Model Architecture

The digital twin of a Smart City comprises the following key components:

- Digital Model: A highly detailed digital model replicates the city's physical infrastructure, public services, and the behaviour of urban systems. This model is continually updated with real-time data to mirror the evolving city environment.
- Federated Learning Framework: The federated learning framework is deployed to facilitate decentralized model training across edge devices. Edge devices include sensors, public transportation systems, and data sources distributed throughout the city.
- Local Models: Each edge device maintains its local model, trained on the data generated by the device. Local models capture insights specific to the area or system they represent.
- Aggregator and Global Model: A central server, acting as an aggregator, receives local model updates from edge devices. It aggregates these updates to compute a global model, representing a collective understanding of the entire city.

5.7.2.3 Challenges Encountered

The implementation of federated learning in a Smart Cities digital twin was met with several challenges:

- Data Diversity: Data from various sensors and systems often vary in format, quality, and update frequency. Harmonizing these diverse data sources for model training was a complex task.
- Data Localization: Different neighbourhoods and areas within a city may have distinct characteristics. Ensuring that the global model can reflect localized insights was a challenge that required careful attention.
- Privacy and Security: Handling citizen feedback data while safeguarding privacy and security was a paramount concern. The deployment of differential privacy and secure aggregation techniques was essential.

5.7.2.4 Successful Outcomes

Despite the challenges, the implementation of federated learning in the Smart Cities' digital twin yielded numerous benefits:

- Real-Time Traffic Management: The digital twin provided real-time traffic insights, enabling dynamic traffic signal adjustments and rerouting to ease congestion and reduce commute times.
- Environmental Monitoring and Resource Optimization: Environmental sensors contributed to real-time air quality monitoring and energy consumption optimization. The city could respond promptly to environmental issues and manage resource utilization efficiently.

- Citizen-Centric Services: Citizen feedback, integrated into the digital twin, allowed the city to tailor public services to the needs and expectations of its residents, enhancing overall urban experiences.
- Data Privacy Compliance: The implementation maintained the privacy of citizen data, ensuring compliance with privacy regulations while benefiting from citizen input.
- Scalability: As the city's infrastructure evolved and new IoT devices were added, the federated learning system seamlessly scaled to incorporate these new data sources.

5.7.3 Challenges in Applying Federated Learning in Digital Twin Contexts

Federated learning, when integrated with digital twins, presents a transformative approach to harnessing decentralized data for modelling and optimization. However, this integration comes with its own set of challenges. Addressing these challenges is crucial for a successful implementation. Some common challenges include:

5.7.3.1 Communication Overhead

The necessity of regular communication between edge devices and a central server in federated learning, specifically in the context of Digital Twins. This communication is essential for exchanging model updates, but it poses a notable challenge, particularly in scenarios that involve a large number of devices or data-intensive model updates. To address this challenge, potential solutions involve implementing techniques such as model compression and quantization. These approaches aim to reduce the size of model updates, thereby improving bandwidth efficiency. Furthermore, optimizing communication protocols and strategically scheduling updates during off-peak hours are recommended strategies for effectively managing and reducing the communication overhead.

5.7.3.2 Synchronization Complexities

Navigating the challenge of synchronizing and aggregating model updates from diverse edge devices in federated learning presents inherent complexities. Variability in update frequencies and processing speeds among devices may lead to synchronization issues. A proposed solution involves implementing asynchronous federated learning frameworks, which allow edge devices to update the central server at their own pace. Additionally, strategies such as weighted aggregation can be employed to harmonize the impact of devices with different update frequencies, aiming for a balanced and effective convergence of the global model. Addressing these intricacies is crucial for the successful implementation of federated learning in scenarios characterized by diverse edge devices.

5.7.3.3 Data Quality Concerns

In digital twin environments, a significant challenge arises from the inherent variability in the quality, accuracy, and reliability of the data they encapsulate. The presence of noisy or erroneous data has a detrimental impact on the training of local models, thereby affecting the global model. To address this challenge, a crucial solution lies in the implementation of robust data preprocessing methodologies. This involves the implementation of techniques such as outlier detection, data cleaning, and error correction. Crucially, implementing data quality checks at the source and integrating feedback loops to iteratively enhance data quality over time emerges as a beneficial strategy, ensuring the integrity and effectiveness of the models in the digital twin ecosystem.

5.7.3.4 Imbalanced Data Distribution

In the context of federated learning, a challenge arises from the uneven distribution of contributions among edge devices to the global model. This imbalance can result from variations in the volume or informational richness of the data generated by different devices, leading to disparate data partitions. To mitigate this challenge, effective strategies need to be implemented. A recommended solution involves incorporating weighted aggregation techniques. By assigning different weights to local model updates based on the quality or contribution of the data from each device, a more balanced and representative aggregation can be achieved. This approach ensures that the valuable insights derived from all participating devices are appropriately considered in the overall updates to the global model, thereby enhancing the overall fairness and effectiveness of the federated learning process.

5.7.3.5 Privacy and Security

In digital twins, a challenge pertains to the handling of sensitive or confidential data, which is a particular concern in sectors such as healthcare or finance. The imperative is to navigate privacy considerations meticulously, balancing the necessity of safeguarding individual information with the collaborative requirements of model training. Addressing this challenge involves implementing privacy-preserving techniques, notably differential privacy and secure aggregation. Differential privacy introduces controlled noise to local updates, thereby safeguarding individual data points from disclosure. Concurrently, secure aggregation plays a crucial role in preserving confidentiality during the aggregation process, ensuring that both raw data and updates are shielded from unauthorized access. These solutions collectively contribute to the establishment of a privacy-conscious framework for digital twins, enabling the effective utilization of sensitive data in a secure and ethically sound manner.

5.7.3.6 Scalability

In the rapidly growing digital twin ecosystems, the key challenge is to ensure the scalability of federated learning methodologies, especially as the number of integrated edge devices continues to increase. Effectively addressing this challenge requires a proactive approach in the design of the system to effectively manage the increasing computational demands. To tackle this, the incorporation of distributed computing frameworks and the utilization of cloud resources emerge as strategic solutions. Leveraging these technologies facilitates the efficient handling of the augmented computational load induced by the growing array of edge devices. Moreover, the implementation of auto-scaling mechanisms stands out as a critical strategy, enabling the system to seamlessly adapt and accommodate the integration of new devices, thereby fortifying the scalability of federated learning within the dynamic milieu of digital twin ecosystems.

5.8 Shaping the Future of Digital Twins and Federated Learning

As we conclude this chapter on employing federated learning in the context of digital twins, it is crucial to look ahead and consider the emerging trends and possibilities that are poised to reshape the landscape of digital twins and federated learning. Several key developments are on the horizon, each of which holds the potential to revolutionize the way digital twins are implemented and leveraged in various domains.

5.8.1 6G Technology and Ultra-Low Latency

The deployment of 6G technology represents a game-changing advancement for digital twins and federated learning. The ultra-low latency, high bandwidth, and network reliability offered by 6G networks are set to unlock real-time capabilities and unprecedented connectivity [95]. This will be particularly impactful in scenarios where rapid data transmission and response times are essential, such as in autonomous vehicles, smart cities, and augmented reality applications. Federated learning will benefit from this by enabling faster model updates and improved coordination among edge devices, contributing to more responsive and agile digital twins [77].

5.8.2 Edge Computing Advancements

Edge computing is set to become a linchpin in the integration of digital twins and federated learning. With increasingly powerful edge devices, the ability to conduct complex AI/ML computations at the edge will grow significantly. This will enable more sophisticated local model training, reducing the reliance on centralized servers for computations. As edge computing evolves, the digital twin experience will become more immersive and responsive, while federated learning will become more distributed and efficient.

5.8.3 Cutting-Edge AI/ML Techniques

Advancements in artificial intelligence and machine learning techniques are continuously expanding the possibilities for digital twins. Techniques like deep learning, reinforcement learning, and generative adversarial networks offer new avenues for modeling and simulating digital twins with greater accuracy and realism. In the context of federated learning, these advanced techniques can further enhance model quality and predictive capabilities. The use of generative models for data augmentation and anomaly detection in digital twins, and the application of reinforcement learning for real-time optimization, are just a glimpse of what's on the horizon.

5.8.4 Hybrid Models and Integration with Physical Systems

The future holds the promise of tightly integrating digital twins with physical systems. Digital twins will become even more "lifelike" by incorporating real-time data from IoT devices, sensors, and connected equipment. This synergy will create new possibilities for predictive maintenance, remote monitoring, and optimizing resource utilization. Hybrid models that combine physical and digital representations will be at the forefront of this evolution, allowing for more accurate simulations and real-time interactions.

5.8.5 Enhanced Interactivity and User Engagement

The evolution of digital twins will bring about more interactive and engaging user experiences. Augmented reality (AR) and virtual reality (VR) technologies, coupled with natural language processing and gesture recognition, will enable users to immerse themselves in the digital twin environment. This will have profound implications for training, education, and collaborative problem-solving. Federated learning will play a pivotal role in personalizing and optimizing these experiences by leveraging insights from distributed edge devices.

5.9 Research Challenges in Employing Federated Learning in Digital Twins

The integration of federated learning with digital twins opens new avenues for research and development, addressing challenges that can significantly advance the field. To stimulate further research, several key challenges need to be addressed:

5.9.1 Model Aggregation Efficiency

Efficient model aggregation is at the core of federated learning, where model updates from various edge devices are combined to create a global model [88]. The challenge here is to design aggregation methods that reduce communication overhead and computational requirements while preserving model quality. Research efforts should focus on developing advanced aggregation techniques, including personalized aggregation based on device performance, differential privacy-enhanced aggregation, and strategies to address imbalanced data contributions, all of which can significantly improve the efficiency of model aggregation in the context of digital twins.

5.9.2 Privacy and Security Mechanisms

Maintaining data privacy and security within federated learning, especially when dealing with sensitive data in digital twins, remains a critical concern [96]. The research challenge lies in advancing privacy-preserving techniques and enhancing security mechanisms. Investigating improvements in differential privacy, secure aggregation, and homomorphic encryption for better data protection is vital. Researchers should also focus on developing methods for detecting and mitigating privacy vulnerabilities in federated learning systems, ensuring that sensitive data remains confidential and secure.

5.9.3 Scalability

As digital twin ecosystems continue to expand, accommodating the increasing number of edge devices and the growing volume of data is essential. Ensuring the scalability of federated learning systems in this context is a significant research challenge [78]. This challenge necessitates exploring distributed computing strategies, cloud-based solutions, and mechanisms for dynamic resource allocation and load balancing to handle the demands of large-scale digital twin environments efficiently.

5.9.4 Data Quality and Preprocessing

Data quality is fundamental in digital twin applications, where inaccurate or noisy data can lead to suboptimal model performance. The research challenge is to develop advanced data cleaning, outlier detection, and data imputation techniques specifically tailored to the nuances of digital twin environments. Research should also investigate strategies for adaptive data quality assessment during model training, enabling the system to handle changing data quality over time effectively.

5.9.5 Heterogeneous Data Sources

Digital twins often integrate data from diverse sources, each with its own format and characteristics. Ensuring that this heterogeneous data is harmonized and effectively utilized presents a significant challenge. Research efforts should be directed toward developing data fusion and transformation techniques that can reconcile the varied data sources. Additionally, exploring adaptive models capable of handling differing data types and structures is essential for improving the utility of digital twin systems.

5.9.6 Real-Time Processing

Many digital twin applications require real-time insights and responsiveness. However, balancing the need for rapid model updates with the challenges of federated learning, which involves communication and aggregation, is a complex problem. Researchers need to develop mechanisms for adaptive communication and aggregation frequency, ensuring that real-time federated learning strategies can provide timely insights while maintaining model accuracy, which is particularly critical in applications like smart cities and manufacturing.

5.9.7 *Explainability and Interpretability*

Federated learning models can be complex and difficult to interpret, making it challenging for users to trust and understand their decisions. Enhancing the explainability and interpretability of federated learning models is a vital research challenge. This involves investigating methods for model transparency, clear explanations of model updates, and making the decision-making process understandable to users. Achieving this is essential for building trust and usability in digital twin applications.

5.9.8 *Interoperability*

Digital twins often need to integrate with existing systems and platforms. Ensuring interoperability between federated learning frameworks and digital twin ecosystems is crucial. Research should focus on investigating interoperable federated learning protocols and standardization efforts that enable seamless integration with various digital twin platforms. This research challenge aims to facilitate the adoption of federated learning in diverse digital twin environments while maintaining compatibility with existing systems.

5.10 Conclusion

This chapter has explored the transformative potential of utilizing federated learning in digital twin applications across various industries. The key takeaways underscore the profound impact of this integration and how it is poised to revolutionize our interactions with digital twins. At its core, federated learning offers a decentralized and privacy-conscious paradigm that enhances digital twins by enabling model training on edge devices. It empowers us to leverage insights from dispersed sources while protecting sensitive data, enabling more intelligent, responsive, and secure digital twin experiences. The chapter has showcased the wide-ranging industry applications, from healthcare and manufacturing to smart cities and agriculture, where the combination of federated learning and digital twins holds immense promise. It has underscored the relevance of this integration in solving real-world challenges and unlocking new opportunities for innovation.

Privacy and security have emerged as central themes, as protecting sensitive data while reaping its benefits remains a priority. Techniques such as differential privacy and secure aggregation are vital for striking a balance between data privacy and utility. Moreover, the chapter has highlighted common challenges, such as communication overhead, synchronization complexities, and data quality concerns, while emphasizing the importance of continuous research and the development of solutions. The advancement of model aggregation efficiency, enhancement of data quality, and development of scalable systems are essential to overcome these challenges.

Looking towards the future, the impacts of 6G technology, advancements in edge computing, and cutting-edge AI/ML techniques are poised to revolutionize digital twins, making them more interactive, responsive, and accurate representations of the physical world. The fusion of these technologies promises a future where the virtual and physical worlds converge, enhancing our interactions with digital twins and revolutionizing how we engage with the physical realm. In conclusion, the integration of federated learning with digital twins represents a groundbreaking approach that has the potential to reshape multiple industries. It empowers us to build smarter, more efficient, and more secure systems while respecting data privacy. As researchers and innovators continue to push boundaries, the future holds the promise of a seamless fusion between the digital and physical realms. This will result in our interactions with digital twins becoming more intelligent and immersive than ever before.

References

1. E. Negri, L. Fumagalli, M. Macchi, A review of the roles of digital twin in CPS-based production systems. Proc. Manufact. **11**, 939–948 (2017)
2. W. Kritzinger, M. Karner, G. Traar, J. Henjes, W. Sihn, Digital twin in manufacturing: a categorical literature review and classification. IFAC-PapersOnline **51**(11), 1016–1022 (2018)
3. U. Tariq, R. Joy, S.-H. Wu, M.A. Mahmood, A. Malik, F. Liou, A state-of-the-art digital factory integrating digital twin for laser additive and subtractive manufacturing processes. Rapid Prototyp. J. (2023). https://doi.org/10.1108/RPJ-03-2023-0113
4. T. Erol, A.F. Mendi, D. Doğan, The digital twin revolution in healthcare, in *2020 4th International Symposium on Multidisciplinary Studies and Innovative Technologies (ISMSIT)* (IEEE, Piscataway, 2020), pp. 1–7
5. T. Sun, X. He, Z. Li, Digital twin in healthcare: recent updates and challenges. Digit. Health **9**, 20552076221149651 (2023)
6. W. Huang, Y. Zhang, W. Zeng, Development and application of digital twin technology for integrated regional energy systems in smart cities. Sustain. Comput. Inf. Syst. **36**, 100781 (2022)
7. N. Fukawa, A. Rindfleisch, Enhancing innovation via the digital twin. J. Product Innov. Manag. **40**, 391–406 (2023)
8. S. Mihai, M. Yaqoob, D. Hung, W. Davis, P. Towakel, M. Raza, M. Karamanoglu, B. Barn, D. Shetve, R. Prasad, H. Venkataraman, R. Trestian, H. Nguyen, Digital twins: a survey on enabling technologies, challenges, trends and future prospects. IEEE Commun. Surv. 1–1 (2022). https://doi.org/10.1109/COMST.2022.3208773
9. A. Bécue, E. Maia, L. Feeken, P. Borchers, I. Praça, A new concept of digital twin supporting optimization and resilience of factories of the future. Appl. Sci. **10**(13), 4482 (2020)
10. D. Zhong, Z. Xia, Y. Zhu, J. Duan, Overview of predictive maintenance based on digital twin technology. Heliyon **9**, e14534 (2023)
11. P. Maheshwari, S. Kamble, A. Belhadi, M. Venkatesh, M.Z. Abedin, Digital twin-driven real-time planning, monitoring, and controlling in food supply chains. Technol. Forecast. Soc. Change **195**, 122799 (2023)
12. A. Vallée, Digital twin for healthcare systems. Front. Digit. Health **5**, 1253050 (2023)

13. D.M. Botín-Sanabria, A.S. Mihaita, R.E. Peimbert-García, M.A. Ramírez-Moreno, R.A. Ramírez-Mendoza, J. de J. Lozoya-Santos, Digital twin technology challenges and applications: a comprehensive review. Remote Sens. **14**(6), 1335 (2022)
14. C. Zhang, Y. Xie, H. Bai, B. Yu, W. Li, Y. Gao, A survey on federated learning. Knowl. Based Syst. **216**, 106775 (2021)
15. S. Abdulrahman, H. Tout, H. Ould-Slimane, A. Mourad, C. Talhi, M. Guizani, A survey on federated learning: the journey from centralized to distributed on-site learning and beyond. IEEE Internet Things J. **8**(7), 5476–5497 (2020)
16. M. Shaheen, M.S. Farooq, T. Umer, B.S. Kim, Applications of federated learning; taxonomy, challenges, and research trends. Electronics **11**(4), 670 (2022)
17. S. Abdulrahman, S. Otoum, O. Bouachir, A. Mourad, Management of digital twin-driven iot using federated learning. IEEE J. Sel. Areas Commun. **41**, 3636–3649 (2023)
18. A. Rizwan, R. Ahmad, A.N. Khan, R. Xu, D.H. Kim, Intelligent digital twin for federated learning in AIoT networks. Internet Things **22**, 100698 (2023)
19. X. Wang, N. Cheng, L. Ma, R. Sun, R. Chai, N. Lu, Digital twin-assisted knowledge distillation framework for heterogeneous federated learning. China Commun. **20**(2), 61–78 (2023)
20. J. Guo, Z. Liu, S. Tian, F. Huang, J. Li, X. Li, K.K. Igorevich, J. Ma, TFL-DT: a trust evaluation scheme for federated learning in digital twin for mobile networks. IEEE J. Sel. Areas Commun. 41, 3548–3560 (2023)
21. M. Jemaa, F. Jemili, A. Charfeddine, M. Msolli, O. Korbaa, Digital twin for a human heart using deep learning and stream processing platforms (2023). https://doi.org/10.21203/rs.3.rs-3359965/v1
22. M. Dietz, G. Pernul, Digital twin: empowering enterprises towards a system-of-systems approach. Bus. Inf. Syst. Eng. **62**, 179–184 (2020)
23. B.A. Talkhestani, T. Jung, B. Lindemann, N. Sahlab, N. Jazdi, W. Schloegl, M. Weyrich, An architecture of an intelligent digital twin in a cyber-physical production system. at-Automatisierungstechnik **67**(9), 762–782 (2019)
24. K. Samir, A. Maffei, M.A. Onori, Real-time asset tracking; a starting point for digital twin implementation in manufacturing. Proc. Cirp **81**, 719–723 (2019)
25. P. Unal, Ö. Albayrak, M. Jomâa, A.J. Berre, Data-driven artificial intelligence and predictive analytics for the maintenance of industrial machinery with hybrid and cognitive digital twins, in *Technologies and Applications for Big Data Value* (Springer, Berlin, 2022), pp. 299–319
26. J. Bao, D. Guo, J. Li, J. Zhang, The modelling and operations for the digital twin in the context of manufacturing. Enterprise Inf. Syst. **13**(4), 534–556 (2019)
27. R. Laubenbacher, J.P. Sluka, J.A. Glazier, Using digital twins in viral infection. Science **371**(6534), 1105–1106 (2021)
28. C. Wu, G. Lorenzo, D.A. Hormuth, E.A.B.F. Lima, K.P. Slavkova, J.C. DiCarlo, J. Virostko, C.M. Phillips, D. Patt, C. Chung, et al., Integrating mechanism-based modeling with biomedical imaging to build practical digital twins for clinical oncology. Biophys. Rev. **3**(2), 021304 (2022)
29. A. Redaelli, E. Votta, Cardiovascular patient-specific modeling: where are we now and what does the future look like? APL Bioeng. **4**(4), 040401 (2020)
30. R. Sahal, S.H. Alsamhi, K.N. Brown, Personal digital twin: a close look into the present and a step towards the future of personalised healthcare industry. Sensors **22**(15), 5918 (2022)
31. I. Voigt, H. Inojosa, A. Dillenseger, R. Haase, K. Akgün, T. Ziemssen, Digital twins for multiple sclerosis. Front. Immunol. **12**, 669811 (2021)
32. S. Elkefi, O. Asan, Digital twins for managing health care systems: rapid literature review. J. Med. Internet Res. **24**(8), e37641 (2022)
33. T. Deng, K. Zhang, Z.J.M. Shen, A systematic review of a digital twin city: a new pattern of urban governance toward smart cities. J. Manag. Sci. Eng. **6**(2), 125–134 (2021)
34. R. Elsehrawy, B. Kumar, R. Watson, A digital twin uses classification system for urban planning & city infrastructure management. J. Inf. Technol. Construct. **26**, 832–862 (2021)
35. A. Martínez-Gutiérrez, J. Díez-González, R. Ferrero-Guillén, P. Verde, R. Álvarez, H. Perez, Digital twin for automatic transportation in industry 4.0. Sensors **21**(10), 3344 (2021)

36. S.A.P. Kumar, R. Madhumathi, P.R. Chelliah, L. Tao, S. Wang, A novel digital twin-centric approach for driver intention prediction and traffic congestion avoidance. J. Reliable Intell. Environ. **4**, 199–209 (2018)
37. C. Hu, W. Fan, E. Zeng, Z. Hang, F. Wang, L. Qi, M.Z.A. Bhuiyan, Digital twin-assisted real-time traffic data prediction method for 5g-enabled internet of vehicles. IEEE Trans. Ind. Inf. **18**(4), 2811–2819 (2021)
38. C. Schwarz, Z. Wang, The role of digital twins in connected and automated vehicles. IEEE Intell. Transp. Syst. Mag. **14**(6), 41–51 (2022)
39. M. Baessler, S. Haufe, H. Meyer, J. Zimdahl, A. Kamtsiuris, R. Meissner, F. Raddatz, Development of a digital twin for aviation research (2020). https://doi.org/10.25967/530329
40. U. Dahmen, J. Rossmann, Experimentable digital twins for a modeling and simulation-based engineering approach, in *2018 IEEE International Systems Engineering Symposium (ISSE)* (IEEE, Piscataway, 2018), pp. 1–8
41. A.F. Mendi, T. Erol, D. Doğan, Digital twin in the military field. IEEE Internet Comput. **26**(5), 33–40 (2021)
42. L. Li, S. Aslam, A. Wileman, S. Perinpanayagam, Digital twin in aerospace industry: a gentle introduction. IEEE Access **10**, 9543–9562 (2021)
43. P. Palensky, M. Cvetkovic, D. Gusain, A. Joseph, Digital twins and their use in future power systems. Digit. Twin **1**, 4 (2022)
44. S.V. Fernandes, D.V. João, B.B. Cardoso, M.A.I. Martins, E.G. Carvalho, Digital twin concept developing on an electrical distribution system—an application case. Energies **15**(8), 2836 (2022)
45. R. Darbali-Zamora, J. Johnson, A. Summers, C.B. Jones, C. Hansen, C. Showalter, State estimation-based distributed energy resource optimization for distribution voltage regulation in telemetry-sparse environments using a real-time digital twin. Energies **14**(3), 774 (2021)
46. N. Bazmohammadi, A. Madary, J.C. Vasquez, H.B. Mohammadi, B. Khan, Y. Wu, J.M. Guerrero, Microgrid digital twins: concepts, applications, and future trends. IEEE Access **10**, 2284–2302 (2021)
47. G. Nasiri, A. Kavousi-Fard, A digital twin-based system to manage the energy hub and enhance the electrical grid resiliency. Machines **11**(3), 392 (2023)
48. R.C. Geyer, T. Klein, M. Nabi, Differentially private federated learning: a client level perspective (2017). arXiv:1712.07557
49. J. Wang, Y. Chen, Federated learning for personalized healthcare, in *Introduction to Transfer Learning: Algorithms and Practice* (Springer, Berlin, 2022), pp. 303–313
50. M.A.P. Putra, S.M. Rachmawati, R.N. Alief, L.A.C. Ahakonye, A. Gohil, D.S. Kim, J.M. Lee, Federated learning-enabled digital twin for smart additive manufacturing industry, in *2023 International Conference on Artificial Intelligence in Information and Communication (ICAIIC)* (IEEE, Piscataway, 2023), pp. 806–811
51. Z. Li, V. Sharma, S.P. Mohanty, Preserving data privacy via federated learning: challenges and solutions. IEEE Consum. Electron. Mag. **9**(3), 8–16 (2020)
52. N. Truong, K. Sun, S. Wang, F. Guitton, Y. Guo, Privacy preservation in federated learning: an insightful survey from the GDPR perspective. Comput. Secur. **110**, 102402 (2021)
53. G. Liu, C. Wang, X. Ma, Y. Yang, Keep your data locally: federated-learning-based data privacy preservation in edge computing. IEEE Netw. **35**(2), 60–66 (2021)
54. W. Luping, W. Wei, L. Bo, CMFL: mitigating communication overhead for federated learning, in *2019 IEEE 39th International Conference on Distributed Computing Systems (ICDCS)* (IEEE, Piscataway, 2019), pp. 954–964
55. X. Wu, X. Yao, C.L. Wang, FedSCR: structure-based communication reduction for federated learning. IEEE Trans. Parallel Distrib. Syst. **32**(7), 1565–1577 (2020)
56. M. Zhang, E. Wei, R. Berry, Faithful edge federated learning: scalability and privacy. IEEE J. Sel. Areas Commun. **39**(12), 3790–3804 (2021)
57. D. Huba, J. Nguyen, K. Malik, R. Zhu, M. Rabbat, A. Yousefpour, C.J. Wu, H. Zhan, P. Ustinov, H. Srinivas, et al., Papaya: practical, private, and scalable federated learning. Proc. Mach. Learn. Syst. **4**, 814–832 (2022)

58. Q. Wu, K. He, X. Chen, Personalized federated learning for intelligent iot applications: a cloud-edge based framework. IEEE Open J. Comput. Soc. **1**, 35–44 (2020)
59. C. Alcaraz, J. Lopez, Digital twin: a comprehensive survey of security threats. IEEE Commun. Surv. Tuts. **24**(3), 1475–1503 (2022)
60. H. Haße, H. van der Valk, N. Weißenberg, B. Otto, Shared digital twins: data sovereignty in logistics networks, in *Data Science and Innovation in Supply Chain Management: How Data Transforms the Value Chain. Proceedings of the Hamburg International Conference of Logistics (HICL)*, vol. 29 (epubli GmbH, Berlin, 2020), pp. 763–795
61. K. Alshammari, T. Beach, Y. Rezgui, Cybersecurity for digital twins in the built environment: current research and future directions. J. Inf. Technol. Construct. **26**, 159–173 (2021)
62. J. Autiosalo, J. Vepsäläinen, R. Viitala, K. Tammi, A feature-based framework for structuring industrial digital twins. IEEE Access **8**, 1193–1208 (2019)
63. C. Angulo, J.A. Ortega, L. Gonzalez-Abril, Towards a healthcare digital twin 1, in *Artificial Intelligence Research and Development* (IOS Press, Amsterdam, 2019), pp. 312–315
64. J. Pang, Y. Huang, Z. Xie, Q. Han, Z. Cai, Realizing the heterogeneity: a self-organized federated learning framework for IoT. IEEE Internet Things J. **8**(5), 3088–3098 (2020)
65. K. Guo, T. Chen, S. Ren, N. Li, M. Hu, J. Kang, Federated learning empowered real-time medical data processing method for smart healthcare. IEEE/ACM Trans. Comput. Biol. Bioinf. (2022)
66. W. Sun, S. Lian, H. Zhang, Y. Zhang, Lightweight digital twin and federated learning with distributed incentive in air-ground 6g networks. IEEE Trans. Netw. Sci. Eng. **10**(3), 1214–1227 (2022)
67. J. Monteiro, J. Barata, M. Veloso, L. Veloso, J. Nunes, A scalable digital twin for vertical farming. J. Ambient Intell. Human. Comput. **14**(10), 13981–13996 (2023)
68. N. Lang, E. Sofer, N. Shlezinger, R.G.L. D'Oliveira, S. El Rouayheb, CPA: compressed private aggregation for scalable federated learning over massive networks, in *ICASSP 2023-2023 IEEE International Conference on Acoustics, Speech and Signal Processing (ICASSP)* (IEEE, Piscataway, 2023), pp. 1–5
69. V. Hegiste, T. Legler, M. Ruskowski, Application of federated machine learning in manufacturing, in *2022 International Conference on Industry 4.0 Technology (I4Tech)* (IEEE, Piscataway, 2022), pp. 1–8
70. B. Han, M.A. Habibi, B. Richerzhagen, K. Schindhelm, F. Zeiger, F. Lamberti, F.G. Prattico̊, K. Upadhya, C. Korovesis, I.P. Belikaidis, et al., Digital twins for industry 4.0 in the 6g era. IEEE Open J. Veh. Technol. (2023)
71. N. Rana, H. Marwaha. Role of federated learning in healthcare systems: a survey. Math. Found. Comput. 7(4), 459–484 (2024). https://doi.org/10.3934/mfc.2023023
72. A. Arora, S. Srivastava, A. Raj, S. Bansal, Appositeness of digital twins in healthcare, in *Digital Twins and Healthcare: Trends, Techniques, and Challenges* (IGI Global, Hershey, 2023), pp. 55–68
73. H. Jeong, T.M. Chung, Security and privacy issues and solutions in federated learning for digital healthcare, in *International Conference on Future Data and Security Engineering* (Springer, Berlin, 2022), pp. 316–331
74. D. Gupta, O. Kayode, S. Bhatt, M. Gupta, A.S. Tosun, Hierarchical federated learning based anomaly detection using digital twins for smart healthcare, in *2021 IEEE 7th International Conference on Collaboration and Internet Computing (CIC)* (IEEE, Piscataway, 2021), pp. 16–25
75. W. Wang, F. He, Y. Li, S. Tang, X. Li, J. Xia, Z. Lv, Data information processing of traffic digital twins in smart cities using edge intelligent federation learning. Inf. Process. Manag. **60**(2), 103171 (2023)
76. W. Piper, H. Sun, J. Jiang, Digital twins for smart cities: case study and visualisation via mixed reality, in *2022 IEEE 96th Vehicular Technology Conference (VTC2022-Fall)* (IEEE, Piscataway, 2022), pp. 1–5

77. M. Saranya, B. Amutha, An insight into federated learning based digital twin applications in smart city, in *2023 5th International Conference on Inventive Research in Computing Applications (ICIRCA)* (IEEE, Piscataway, 2023), pp. 1295–1302
78. S.P. Ramu, P. Boopalan, Q.V. Pham, P.K.R. Maddikunta, T. Huynh-The, M. Alazab, T.T. Nguyen, T.R. Gadekallu, Federated learning enabled digital twins for smart cities: concepts, recent advances, and future directions. Sustain. Cities Soc. **79**, 103663 (2022)
79. D. Wang, N. Zhang, M. Tao, Adaptive clustering-based model aggregation for federated learning with imbalanced data, in *2021 IEEE 22nd International Workshop on Signal Processing Advances in Wireless Communications (SPAWC)* (IEEE, Piscataway, 2021), pp. 591–595
80. Y. Zhou, Q. Ye, J. Lv, Communication-efficient federated learning with compensated overlap-fedavg. IEEE Trans. Parallel Distrib. Syst. **33**(1), 192–205 (2021)
81. H. Fereidooni, S. Marchal, M. Miettinen, A. Mirhoseini, H. Möllering, T.D. Nguyen, P. Rieger, A.R. Sadeghi, T. Schneider, H. Yalame, et al., Safelearn: secure aggregation for private federated learning, in *2021 IEEE Security and Privacy Workshops (SPW)* (IEEE, Piscataway, 2021), pp. 56–62
82. H. Zhu, On the relationship between (secure) multi-party computation and (secure) federated learning (2020). arXiv:2008.02609
83. E. Hosseini, A. Khisti, Secure aggregation in federated learning via multiparty homomorphic encryption, in *2021 IEEE Globecom Workshops (GC Wkshps)* (IEEE, Piscataway, 2021), pp. 1–6
84. Y. Luo, W. Luo, R. Zhang, H. Zhang, Y. Shi, Robust peer-to-peer learning via secure multi-party computation. J. Inf. Intell. (2023). https://doi.org/1.10.1016/j.jiixd.2023.08.003
85. J. Ma, S.A. Naas, S. Sigg, X. Lyu, Privacy-preserving federated learning based on multi-key homomorphic encryption. Int. J. Intell. Syst. **37**(9), 5880–5901 (2022)
86. B. Chen, A. Bakhshi, G. Batista, B. Ng, T.J. Chin, Update compression for deep neural networks on the edge, in *Proceedings of the IEEE/CVF Conference on Computer Vision and Pattern Recognition* (2022), pp. 3076–3086
87. W. Xu, W. Fang, Y. Ding, M. Zou, N. Xiong, Accelerating federated learning for IoT in big data analytics with pruning, quantization and selective updating. IEEE Access **9**, 38457–38466 (2021)
88. K. Wei, J. Li, M. Ding, C. Ma, H.H. Yang, F. Farokhi, S. Jin, T.Q.S. Quek, H.V. Poor, Federated learning with differential privacy: algorithms and performance analysis. IEEE Trans. Inf. Forensics Secur. **15**, 3454–3469 (2020)
89. K. Wei, J. Li, M. Ding, C. Ma, H. Su, B. Zhang, H.V. Poor, User-level privacy-preserving federated learning: analysis and performance optimization. IEEE Trans. Mobile Comput. **21**(9), 3388–3401 (2021)
90. I. Errandonea, S. Beltrán, S. Arrizabalaga, Digital twin for maintenance: a literature review. Comput. Ind. **123**, 103316 (2020)
91. W. Sun, S. Lei, L. Wang, Z. Liu, Y. Zhang, Adaptive federated learning and digital twin for industrial internet of things. IEEE Trans. Ind. Inf. **17**(8), 5605–5614 (2020)
92. T. Kalsoom, N. Ramzan, S. Ahmed, M. Ur-Rehman, Advances in sensor technologies in the era of smart factory and industry 4.0. Sensors **20**(23), 6783 (2020)
93. M. Shaheen, M.S. Farooq, T. Umer, Reduction in data imbalance for client-side training in federated learning for the prediction of stock market prices. J. Sens. Actuat. Netw. **13**(1), 1 (2024)
94. I. Kevin, K. Wang, X. Zhou, W. Liang, Z. Yan, J. She, Federated transfer learning based cross-domain prediction for smart manufacturing. IEEE Trans. Ind. Inf. **18**(6), 4088–4096 (2021)
95. R. Gupta, D. Reebadiya, S. Tanwar, 6g-enabled edge intelligence for ultra-reliable low latency applications: vision and mission. Comput. Standards Interfaces **77**, 103521 (2021)
96. D. Chen, D. Wang, Y. Zhu, Z. Han, Digital twin for federated analytics using a bayesian approach. IEEE Internet Things J. **8**(22), 16301–16312 (2021)

Chapter 6
Security Attacks in Digital Twin-Enabled Wireless System

Muhammad Amjad

6.1 Introduction

The futuristic applications of Internet of Everything (IoE) such as smart cities, tactile Internet, haptic communications, and extended reality have stringent requirements for security, latency, throughput, connectivity, and reliability [1–3]. To meet these stringent requirements for the 5G and beyond 5G networks use cases, a new framework is required. This framework can then be used to troubleshoot, manage, and optimize the underlying use cases of 5G and beyond 5G networks and IoE applications [4–7]. This could be achieved with the concept of digital twins. Digital twins are the virtualization of the physical 5G and beyond 5G processes with their respective communications protocols and paradigms (algorithms) coupled with security and privacy measures. Through digital twins, wireless networks can be made (i) online-learning systems and (ii) self-configured systems [6]. The advanced wireless networks with these features can meet the stringent requirements of security, massive connectivity, higher throughput, low latency, and reliability. These networks could also be managed, optimized, and troubleshot via the employment of digital twins.

Self-configured and online learning systems can be implemented using digital twins to secure the advanced wireless systems [6]. A digital twin, which is the virtualization of the physical systems, usually takes the inputs from the physical systems and then manage or optimize the systems [8]. The impact of different security threats or vulnerabilities can be tested while utilizing the real data from the physical systems and then the security solutions and measures (patching) can be tested using the digital twins [9]. However, like other physical or virtual systems,

---

M. Amjad (✉)
School of Engineering, Computing and Mathematical Sciences, University of Wolverhampton, Wolverhampton, UK
e-mail: m.amjad6@wlv.ac.uk

M. K. Afzal et al. (eds.), *Digital Twins for Wireless Networks*,
https://doi.org/10.1007/978-3-031-73679-7_6

wireless-enabled digital twins are also at risk of different security risks such as malicious data and jamming [8].

Extensive work has been done on the digital twins-enabled wireless networks. Digital twin networks have been discussed in [10]. This work explained in detail the evolution of digital twins and then the emergence of the digital twin networks. Different communication modes such as physical-to-physical (P2P), physical-to-virtual (P2V), and virtual-to-virtual (V2V) in digital twin networks were explained in detail. The challenging metrics such as latency, reliability, and connectivity were also compared for the P2P, P2V, and V2V modes in digital twin networks. Intelligent reconfigurable surfaces (IRS) are the major player for the beyond 5G networks such as 6G. Digital twin for the IRS was discussed in [11]. A new paradigm called environmental twin-assisted communications was introduced to manage and optimize the IRS for the use cases of 6G networks. This framework was based on a deep learning model that produced a self learning environment for the IRS.

Different security attacks with their potential solutions in digital twins-enabled wireless networks were investigated in [8]. In this work, an overview of different security threats and concerns in physical space and twin space was provided. A secure digital twin communications model for the wireless systems was also studied with different layers of communication. Two use cases related to security concerns such as anti-jamming and label flipping attack were investigated with their potential defense strategies. Some work has also been done considering the security attacks and their defense in practical applications. For example, the impact of the security attacks on the microgrids was investigated using the digital twins [12]. In this work, an IoT-based digital twin of the cyber-physical system was investigated for the microgrids. This digital twin-based framework can counter the coordinated attacks on the system.

Existing works take into consideration the different security attacks and provide a framework to mitigate these attacks using digital twin technology. Motivated by these works, in this chapter we provide an overview of the different security attacks in physical as well as digital twins-enabled wireless networks. Considering the layering approach of the communications model, different security attacks on different layers of the physical and twin layers are discussed in detail. Also, the security measures to mitigate the impact of these attacks in both layers (physical and twin) are highlighted. The open issues, challenges and future research directions regarding the potential security attacks in digital twins-enabled wireless networks with their solutions are discussed along with their higher-level insights.

6.2 Overview of Wireless Enabled Digital Twin

Digital twin is a digital version of the physical systems or processes that takes information as input from the physical systems (real world) and then makes decisions or controls the system [6]. Any security breach in this digital twin based wireless networks could put the system in a compromised position, and futuristic

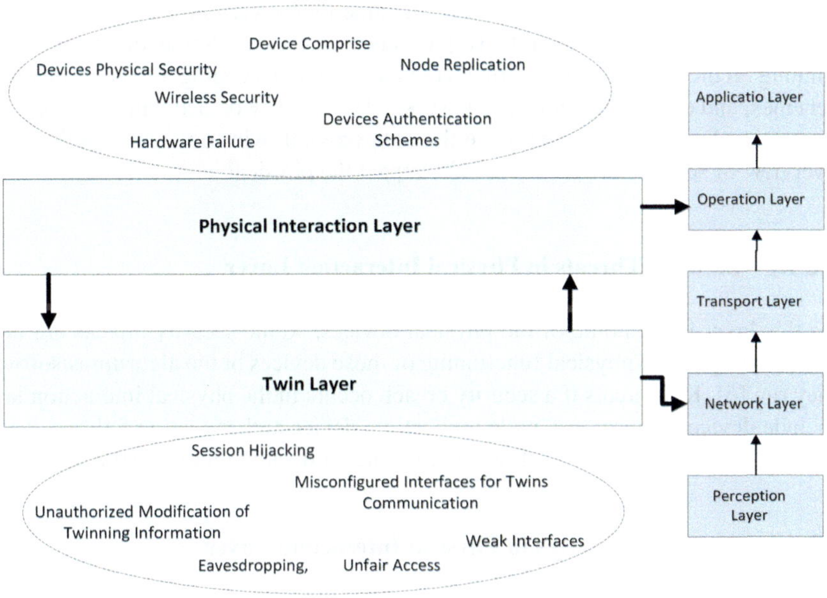

Fig. 6.1 Architecture for digital twins for wireless networks with the potential security attacks

applications could be impacted. An architecture of the digital twin for the wireless networks with the security threats has been shown in Fig. 6.1 [6]. This architecture consists of the physical interaction layer and the twin layer. A third layer, the service layer, could also be part of this architecture. However, this third layer is outside the scope of this book chapter. The physical interaction layer comprises all the physical systems such as wireless devices, connectors, and servers. The twin layer is the logical or virtual layer that replicates the physical layer with algorithms and software. The third layer is the service layer with end services such as smart grids, smart cities, and other futuristic applications.

6.2.1 Physical Interaction Layer

Digital twins for wireless networks have gained much attention in industry and academia. The architecture of the wireless-enabled digital twins consists of two layers: the physical interaction layer and the twin layer. Both layers interact with one another for the seamless wireless activity to achieve higher throughput, low latency, and higher reliability [13]. The physical interaction layer is the base layer for digital twin technology. This layer needs to be secured from potential security attacks as it hosts physical devices such as routers, switches, servers, and other smart devices. There are some basic interfaces required to achieve interaction between the physical and twin layers. Some of the interfaces include twin layer to twin layer, twin layer to physical layer, twin layer to service layer interfaces. Some of the processes running on the physical interaction layer include channel coding, resource allocation schemes, and communications frameworks. Any security breach within the physical interaction layer could compromise the entire digital twin framework. Below is an overview of security attacks that could impact the physical interaction layer.

6.2.1.1 Security Threats in Physical Interaction Layer

As this layer is the home of the physical devices, so the security attacks are more likely to impact the physical functioning of those devices or the algorithms/software they run [6]. Key threats if a security breach occurs in the physical interaction layer include device compromise, node replication, device authentication failures, device physical security attacks, wireless security, and hardware failures are the key threats.

6.2.1.2 Security Measures in Physical Interaction Layer

Effective security solutions in response to the security attacks are required to maintain seamless wireless connectivity in digital twins for wireless networks. Security measures can be applied either to the physical devices or at the interface level. Some of the accepted security measures at the physical interaction layer include secured device authentication protocol, effective wireless encryption schemes, and secure transmission schemes [14].

6.2.2 Twin Layer

The architecture for digital twin in wireless networks has a twin layer that hosts various algorithms and software to control and make decisions. The twin layer uses different modeling approaches to model physical systems/processes. These modeling approaches can include data-driven modeling, 3D modeling, and mathematical modeling. To achieve seamless wireless connectivity in digital twin-enabled wireless networks, the twin layer should be resilient against various security threats. Below are details of different security attacks on the twin layer along with their potential solutions [6, 8].

6.2.2.1 Security Threats in Twin Layer

The twin layer models the physical object using various modeling schemes. It is always susceptible to different security attacks. Some of the security concerns for the twin layer include session hijacking, unauthorized modification of twinning information, eavesdropping, unfair access, and weak interfaces [15].

6.2.2.2 Security Measures in Twin Layer

Security measures have been introduced at the twin layer to mitigate the impact of different security attacks on wireless-enabled digital twins networks. Security threats can be addressed by properly encrypting twin control information. Data theft and leakage can also be prevented at the twin layer to protect against various security attacks [16].

6.3 Security Threats in Digital Twin enabled Wireless Networks

Digital twin-enabled wireless networks combine physical systems with digital/virtual systems. Any security breach could impact both systems and lead to interruptions in wireless connectivity. In this section, we discuss the impact of different security attacks on digital twin-enabled wireless networks, considering the different layers of the communications model [8]. The layered architecture provides an in-depth understanding of different security attacks and their impact on transmission systems. Figure 6.2 illustrates the impact of various security attacks on different layers of the communications model. Below are the details of each layer with different security issues in case of security breaches in the system.

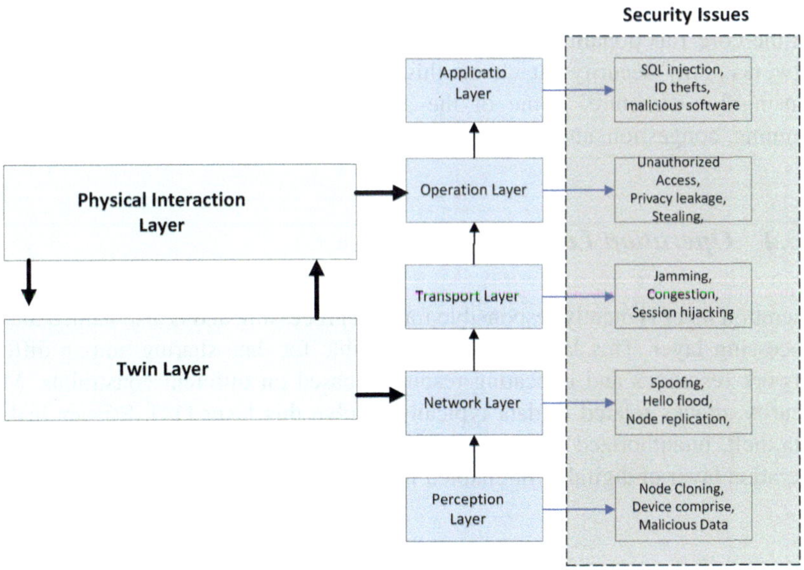

Fig. 6.2 Layered Architecture showing different security attacks on the digital twin enabled wireless networks

6.3.1 Perception Layer

Perception layer, also called the sensor layer, is responsible for collecting data from the physical world [14]. It is the layer where all sensors, actuators, and servers reside. The main security concerns at the perception layer include data tampering, node cloning, passive interface attacks, tag destruction attacks, and tag removing attacks. Most commonly, attacks at the perception layer involve tag attacks and relay attacks.

6.3.2 Network Layer

Network layer, also called the transmission/routing layer in the communications paradigm, acts as a relay between the perception layer and the application layer. Any security attack on this layer could impact the functionality of the whole communication paradigm. Few security attacks that could occur at the network layer include hello flood, spoofing, node replication, cloning attacks, network attack, and man-in-the-middle attacks [8, 14].

6.3.3 Transport Layer

Reliable communications, congestion avoidance, and guaranteeing packet delivery are the core functionalities of the transport layer in digital-twin enabled wireless networks. Any security attack on this layer could result in the breakdown of transmission structure. Some of the attacks related to the transport layer are jamming, congestion, and session hijacking [8].

6.3.4 Operation Layer

Operation layer is mainly responsible for data processing and is also named the data processing layer. This layer is also responsible for data sharing among different network resources and allocating resources based on different constraints. Major security attacks related to data typically involve this layer [14]. Privacy leakage, data theft, unauthorized access, and stealing are the main security concerns at the operation layer of digital twin-enabled wireless networks.

6.3.5 Application Layer

This layer is responsible for delivering final messages of communications and handles the messaging protocols. Any security breach at this layer could hamper services and running applications. Some application layer security attacks include software intersection, SQL injection, cross site scripting, and ID theft [8].

6.4 Suggested Security Measures in Digital Twin Enabled Wireless Networks

Security the digital-twin enabled wireless transmission from different security attacks is the core work of cyber security researchers and engineers. Different measures are adapted to secure the networks from internal and external security attacks. Figure 6.3 shows the different security measures that have been used to protect the digital twin enabled wireless networks at different layers of the physical interaction layer and the twin layer.

6.4.1 Perception Layer

Perception layer as mentioned above is the home of the physical devices such as sensors, servers, and actuators. The security measures at this layer are related to the protection of the physical devices such as device security, preventing data leakage,

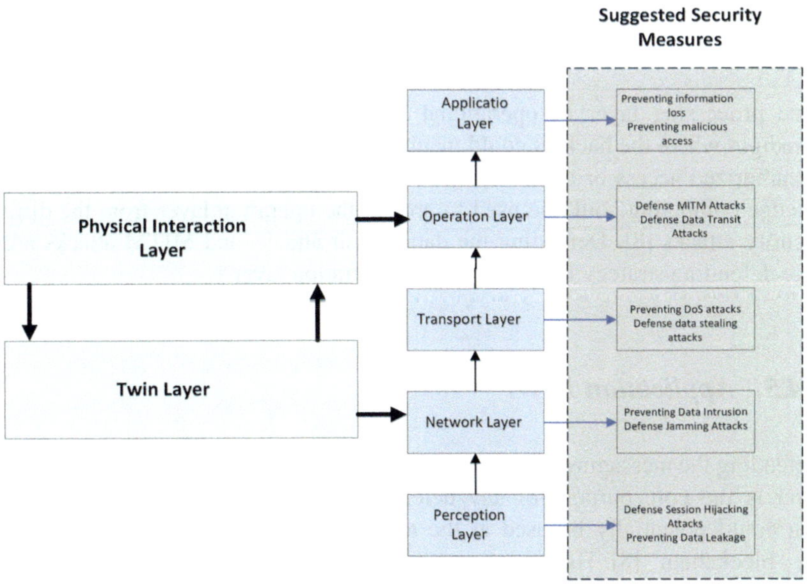

Fig. 6.3 Layered architecture showing potential security solutions for the digital twin enabled wireless networks

and defense session hijacking attacks. The tools that could be useful for defending this layer is the cryptography [8].

6.4.2 Network Layer

Transmission layer or the network layer deals with the data. Therefore, the security solutions to avoid the data leakage, data intrusion, defending jamming attacks, and preventing the unauthorized channel access are mandatory to protect the routing layer. The tool that could be beneficial in protecting this layer could be game theory [8].

6.4.3 Transport Layer

Securing the transport layer is the to ensure the timely delivery of the transmission packets. Various defense strategies are accepted as a tool to defend the transport layer. For example, defending the DoS attacks, preventing node replication attacks, and defending the data stealing attacks are few to name, the security measures applied at the transport layer [14].

6.4.4 Operation Layer

Data processing layer or operational layer is the part of the communications paradigm where the hackers could manipulate the users data or system data to gain unauthorized access or control of the digital-twin enabled wireless networks. Fog or edge computing could be used to protect the operation layer from the different security attacks [8]. Defending the data transit attacks and MITM attacks are the core defending strategy for protecting the operation layer.

6.4.5 Application Layer

Defending the messaging protocols from different security attacks at the application layer is the core purpose of any defense strategy. One of the core technology that could potentially be used at the application layer against security threat is the "blockchain" [8]. Having a good defense against spoofing attacks, preventing malicious access, and avoiding information loss are the key defending strategies from security attacks at the application layer.

6.5 Open Issues, Challenges, and Future Research Directions

Different security attacks in digital twin enabled wireless networks have been presented here. Digital twin enabled wireless networks has been investigated here while taking into consideration the different security threats. However, it is important to investigate the potential impact of the different security attack as a case study with its detection to attack defense. At each layer different security attacks have been investigated, however the cross layer security attacks still need attention. Cross-layer security attacks pose a serious threat to achieve the seamless connectivity in digital twin enabled wireless networks. Digital twin for wireless communications is the combination of physical interaction layer, twin layer, and service layer. Existing works on cyber security considers the security attacks detection and their defense at the physical interaction layer and twin layer. However, still there is need for the investigation of different potential security attacks at the service layer.

6.6 Conclusion

In this chapter, different security attacks at the digital-twin enabled wireless networks have been overviewed considering in detail. Two layer architecture of the digital-twin enabled wireless networks namely, physical interaction layer and twin layer is also investigated while considering the different potential security attacks at these layers. Most specifically, the communication paradigm of the digital twin enabled wireless networks has been analysed using the different layers. Different security attacks on these layers have been investigated in detail with the key functionality of each layer. Also, the security measures or solutions for different security attack at each layer were presented while considering the twin layer and the physical interaction layer.

References

1. M. Amjad, L. Musavian, S. Aïssa, NOMA versus OMA in finite blocklength regime: link-layer rate performance. IEEE Trans. Veh. Technol. **69**(12), 16253–16257 (2020)
2. M. Amjad, L. Musavian, M.H. Rehmani, Effective capacity in wireless networks: a comprehensive survey. IEEE Commun. Surv. Tuts. **21**(4), 3007–3038 (2019)
3. M. Amjad, F. Akhtar, M.H. Rehmani, M. Reisslein, T. Umer, Full-duplex communication in cognitive radio networks: a survey. IEEE Commun. Surv. Tuts. **19**(4), 2158–2191 (2017)
4. L.U. Khan, W. Saad, D. Niyato, Z. Han, C.S. Hong, Digital-twin-enabled 6G: vision, architectural trends, and future directions. IEEE Commun. Mag. **60**(1), 74–80 (2022)
5. L. Bariah, H. Sari, M. Debbah, Digital twin-empowered communications: a new frontier of wireless networks. IEEE Commun. Mag. **61**(12), 24–36 (2023)
6. L.U. Khan, Z. Han, W. Saad, E. Hossain, M. Guizani, C.S. Hong, Digital twin of wireless systems: overview, taxonomy, challenges, and opportunities. IEEE Commun. Surv. Tuts. **24**(4), 2230–2254 (2022)

7. P. Almasan, M. Ferriol-Galmés, J. Paillisse, J. Suárez-Varela, D. Perino, D. López, A.A.P. Perales, et al., Network digital twin: context, enabling technologies, and opportunities. IEEE Commun. Mag. **60**(11), 22–27 (2022)
8. W. Wang, Y. Yang, L. U. Khan, D. Niyato, Z. Han, M. Guizani, Digital Twin for Wireless Networks: Security Attacks and Solutions, in IEEE Wireless Communications, **31**(3), 278–285 (2024)
9. S. Suhail, R. Jurdak, R. Hussain, Security attacks and solutions for digital twins (2022). arXiv:2202.12501
10. Y. Wu, K. Zhang, Y. Zhang, Digital twin networks: a survey. IEEE Internet Things J. **8**(18), 13789–13804 (2021)
11. B. Sheen, J. Yang, X. Feng, M.M.U. Chowdhury, A digital twin for reconfigurable intelligent surface assisted wireless communication (2020). arXiv:2009.00454
12. A. Saad, S. Faddel, T. Youssef, O.A. Mohammed, On the implementation of IoT-based digital twin for networked microgrids resiliency against cyber attacks. IEEE Trans. Smart Grid **11**(6), 5138–5150 (2020)
13. F. Tao, H. Zhang, A. Liu, A.Y.C. Nee, Digital twin in industry: state-of-the-art. IEEE Trans. Ind. Inf. **15**(4), 2405–2415 (2018)
14. H.A. Khattak, M.A. Shah, S. Khan, I. Ali, M. Imran, Perception layer security in Internet of Things. Future Gen. Comput. Syst. **100**(2019), 144–164 (2019)
15. J. Jagannath, K. Ramezanpour, A. Jagannath, Digital twin virtualization with machine learning for IoT and beyond 5G networks: research directions for security and optimal control, in *Proceedings of the 2022 ACM Workshop on Wireless Security and Machine Learning* (2022), pp. 81–86
16. Y. Xing, Cyber security digital twin simulations for wireless networks. Ph.D Dissertation, University of British Columbia (2023)

Chapter 7
Metaverse Enabled UAV in Disaster Scenario

Rooha Masroor, Muhammad Khalil Afzal, and Muhammad Naeem

7.1 Background

History is marked with various heart-rending disaster events where whole civilizations were affected instantly. Information dissemination and providing help in such scenarios are crucial. Disasters have affected humans from the dawn of human life in various forms like earthquakes, floods, tornadoes, and many more. Researchers and organizations kept on working for the management of disasters and provided many solutions to save human life, property, and country prosperity. This management helps in saving human life up to an extent but is unable to provide some proper solution where a human or some aid can be released urgently to get in connection with people. Conclusively, Information dissemination is crucial and should be two-way, i.e., from affected areas to saving authority and from saving authority to an affected area is important. Recently, flood disasters in Pakistan and earthquakes in Turkey and Syria hit so hard that millions of people were affected not in terms of property loss only but also health and lives. According to the World Bank, 33 million were affected by flood disasters in Pakistan with 1730 precious human losses. In recent earthquakes in Turkey, 42,310 people lost their lives and

R. Masroor (✉)
Department of Electrical Engineering and Technology, Muhammad Nawaz Sharif University of Engineering and Technology, Multan, Pakistan

M. K. Afzal
Department of Computer Science, COMSATS University Islamabad, Wah Campus, Islamabad, Pakistan
e-mail: khalilafzal@ciitwah.edu.pk

M. Naeem
Department of Electrical and Computer Engineering, COMSATS University Islamabad, Wah Campus, Punjab, Pakistan
e-mail: mnaeem@ciitwah.edu.pk

M. K. Afzal et al. (eds.), *Digital Twins for Wireless Networks*,
https://doi.org/10.1007/978-3-031-73679-7_7

108,368 people were injured. Damage surveys indicate that 173,000 buildings in 11 provinces collapsed or are heavily damaged. In post-disaster situations, the road links to the affected sites were blocked due to massive destruction. Emergency communication links and medical aid maintenance are difficult in these areas. In such situations, Metaverse with an infrastructure-less technology,i.e., UAV is an ideal solution; where the mirror image of the disastrous area will make a bridge between the helping organizations and the affected people.

7.2 Metaverse and Digital Twin Concepts

The word metaverse originates from two words meta and verse. Meta means transcendence and world means universe. The metaverse translates the world into the virtual world with all the possibilities that are not possible in the physical world. The metaverse is a translation of the real world into the digital world with all the exceptional features of the upcoming modern era, i.e., an environment of the internet of things with virtual space that connects people far away for social meetings, meta-economics, and many more. To understand the concept in depth and clarity, the emerging technology is broken up into four main parts which will further elaborate the details, first is the metaverse element, second is metaverse components, third is metaverse communication technology, and fourth is metaverse-enabling technology. Figure 7.1 is a pictorial view of the following description with sub-categorization of each concept.

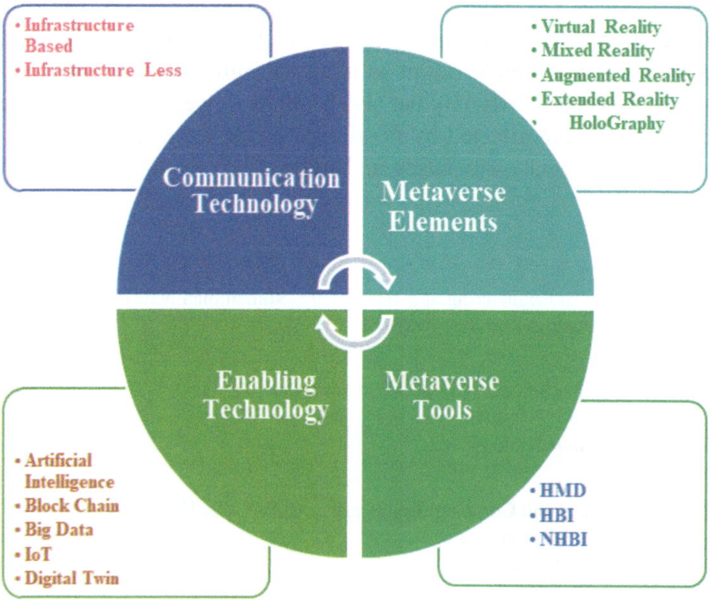

Fig. 7.1 Metaverse concepts

7.2.1 *Metaverse Elements*

Elements of metaverse describe how metaverse technology passed through the evolutionary stage from the physical world to the digital world and reached a stage where it is difficult for someone to recognize or differentiate between the real or imaginary world. Following are the evolutionary stages and sub-classifications of metaverse elements.

- **Augmented Reality:** In this categorization, the real and physical worlds both are approachable and differentiable. A user can intervene both in the physical and virtual worlds. This can be exemplified with the help of the Pokemon game where the camera enhancement feature is used to depict the 3D features and illustrate how the real and virtual worlds are interacting. For example tablets and smartphones.
- **Virtual Reality:** This category can be best explained with the headsets mounted on the human faces where a user is unable to differentiate between the real and virtual world. The virtual world is completely fictional and the real world is in shutdown mode.
- **Mixed Reality:** Collaboration of the real and virtual world is depicted as mixed reality. This will map the virtual world to the real world. This terminology can be better understood with the help of a furniture set where presenting in the real-world wearing metaverse components, a person can decorate a home by placing things according to dimensions and requirements.
- **Extended Reality:** Extended reality will cover all aspects including AR, VR, and MR. Although AR and VR include some aspects of metaverse concepts but ER shades all the elements of ER.
- **Holography:** In holography, a human doesn't need any headset to wear. It is a pre-recorded 3D image that can be later used in multiple applications like virtual meetings.

7.2.2 *Metaverse Components*

Metaverse components play an important role in deploying metaverse technology since it is the hardware part and provide interaction between the metaverse world and the human visionary world. It assists the human eye and ear in visualizing the virtual world with the help of a Head-mounted display (HMD), Hand based input device (HBI), Non-hand based input device (NHBI).

- **HMD:** HMD founds its popularity in major applications, e.g., military, medical sciences, and engineering, and have now found scope in the metaverse world and is renowned as its core hardware component where the user will get completely immersed in the virtual world through position and orientation. However, the change of view in the real world delivers the same change of view in the virtual world through the movement of the screen. These devices are advantageous due to cost and space but are found disadvantageous in terms of latency.

- **HBI:** HBI devices based on haptic technology where active and passive haptic types are further used. Active haptic is for creating virtual pressure while passive haptic is for physical tactile. Based on the meta technology, haptic gloves are used that help in the realization of physical objects (e.g., IKEA home decor) where one can adjust the things by placing them with hand haptic. Active haptic helps improve changes according to human response while passive haptic is more relatable with human presence.
- **NHBI:** These devices works on the principle of eye tracking, head tracking and voice input device, and, many more. In eye tracking and head tracking only eyes and head move and the response will be calculated accordingly to the gesture movements renowned in phobias. On the other hand, voice input devices are exemplary for long text messages and conversations with the help of a virtual keyboard. This category can aid medical sciences where psychological patients got paralyzed due to some chronic fears.
- **Motion-based input:** This device is used to provide realism to users in the metaverse. Here, the components of the metaverse use the physical sense of space or gravity,body tracking and treadmill are used to provide accurate motion information. They are based on two technologies active and passive technology. In the former, there is appropriate feedback based on the users' behavior while in the latter there is a fixed scenario where a sense is delivered to the user.

7.2.3 Metaverse Communication Technologies

Metaverse core methodology requires all-time internet connectivity because the technology is based on communication between physical and imaginary worlds. This uninterrupted way of communication can be done in two ways.

- **Infrastructure Based** Infrastructure-based technologies include cellular infrastructure i.e., traditional conventional setup and wireless technologies which can assist metaverse. 5G and beyond are finding major servers in metaverse.
- **Infrastructure Less** Tackle large volumes of data that need all-time communication since digital twin, hologram needs big backhauls and clouds to support them. A dedicated structure can help maintain cloud and backhaul systems. Therefore, an alternative solution can be UAVs which have already gained much attention from multiple perspectives. Over the last few decades UAVs have been known by different names, i.e., drones, remotely operated pilots operate the system with momentous attention in different domains of civilian [1], industrial [2], and military [3] due to its maneuverability, stability. and endurance. Dilation in applications of UAVs has been found exceptionally due to their faster evolution in device-to-device communication, mobile edge computing, green internet of things (GIoT) [4], 5G [5], and B5G [6]. Similarly, in this chapter UAVs are used

with the concept of metaverse. UAVs applications are segregated according to operational control of UAV; can be operated without a pilot and with a pilot. Frequently, UAVs gain popularity with pilot-less aircraft function that allows the UAV to fly without requiring any onboard operator. While UAVs with pilots need an operator on the ground that will remotely control the situation depending on the scenario requirement. Herein, we will equip our model with a remotely involved pilot on the ground distant from a disastrous site working with a metaverse server using Digital Twin technology.

7.2.4 Metaverse Enabling Technologies

Following are the enabling technologies that support the concept of metaverse. We have briefly described all the technologies.

- **BlockChain:** Metaverse is all about the connection of multiple platforms together, that need a supportive system to gather information and hold them together. Using a centralized method is not a fruitful solution, because the amount of data is very large, and privacy and transparency of data are also issues to tackle. Hence, blockchain paved its way in metaverse which is a decentralized method through which data will be stored in blocks rather than any structured form. The information containing block will be next connected to another block and the chain of blocks will be formed. The blockchain mechanism works for data storage, data acquisition, data translation, data interoperability, and data privacy. The aforementioned terms will help users get rid of a centralized system where users need to talk
- **Artificial Intelligence** Machine learning, later advanced artificial intelligence makes decision and carry out tasks without human intervention. This field can contribute to metaverse technology in different applications popularly in medicine through 3D immersive experience. The counts of patients with multiple diseases and multiple doctors are in large numbers, AI with the help of metaverse can help in assisting doctors by providing electronic health records. The collaboration of metaverse and AI will not only help in analyzing the data but also help in forecasting of increase in diseases and emergency responses.
- **Internet of Things** Internet of Things (IoT) is the combination of the human and machine world where one can perform any task probably without human interaction. The involvement of low-cost sensors and chips made the availability of IoT to almost every work-life application whether for domestic or industrial purposes. The integration of the metaverse concept in IoT i.e., the combination of physical and digital world will help civil industry, medicine field and offices to take decisions and implementation of them in real world through metaverse. To clarify, let us exemplify with an office where different devices are working properly and connected through IoT while metaverse will help in commanding the different IoT tasks without being present there.

- **Digital Twin** Digital twin concept found its roots back in 1991 by author David Gelernter which later enhanced by NASA in 2010. Digital twin is the replica of physical world where everything will have is virtual image in an imaginary world. Digital twin is the building block of metaverse technology. It is helpful in multiple scenarios, e.g. industry, medicine, offices, and, many more. A virtual image of the physical structure will be present to assess the requirements and lack of any task. For civil construction projects, digital twins will help in analyzing the site area, the manpower requirement, and the forecasting of any damage.
- **Big Data** Big data is renowned with three terms Volume, variety and, velocity. Collection of data from various sources at high velocities and increased volume is Big Data. Big Data is a key term in metaverse since it collects data in large volumes. Collaboration of metaverse and big data will help industries and organizations to get a single platform cloud that can be used for analytical purposes.

Literature Review Machine learning is core of digital twin technology [7]. Complex data-driven models are made with a network digital twin concept to run in real time. A case study is elaborated with some baseline methods which are later compared with QoS routing optimization method to show the network performance. The authors in [8] have proposed a dual objective optimization problem whose goals are to reduce the de-synchronization time between CT and reality and minimize the summation of loss in encountered episodes. A digital twin example is described in [9] with the help of five different implementation layers including visualization, synchronization, modeling and simulation, federated digital twin, and intelligent Digital twin services. These layers will help in the present and future deployment of digital technology in any upgraded application. Technologies and elements involved in the implementation of different layers are also discussed. A dynamic placement model of digital twin has been presented in [10] with social capabilities that lie under the tree of IoT. The proposed system is modeled with a quadratic assignment problem which is later resolved with approximation method and two relaxation methods are also introduced to reduce the computational complexity of the proposed solution. Computational offloading, service caching, and latency are the main issues in making digital twin concept possible in any implementation of real-life application [11, 12]. The authors have designed the computational problem into a Mixed integer non-linear program and resolved it with the help of decision theory. The decision theory will help in making a decision table that analyses the system's computational offloading and caching. The end-to-end latency with Qos constraints and computation resources at IoT devices is resolved with an iterative optimizing algorithm in [12]. Another researcher has resolved the problem of low-latency edge computing by objectifying various network properties. The goal is bifurcated into two twin placement and twin migration problems. To achieve an optimal solution deep reinforcement learning algorithm is proposed. The authors in [13] have minimized the latency by keeping the goals of optimising power, user association, offloading, power to transmit, and users' rate. The system is subjected to set of non-convex constraints which are resolved with inner approximation and

Table 7.1 Comparison to existential research. T=Technical, S=Survey, HM=Head Mounted, HBI=Hand based input, NHBI=Non-hand based input

Ref.	Year	Review type		Technology type			Metaverse components			Metaverse elements		
		T	S	5G	6G	UAV	HM	HBI	NHBI	AR	VR	MR/ER
[16]	2022		✓		✓					✓		
[17]	2022	✓									✓	
[18]	2022		✓				✓	✓	✓			✓
[19]	2022		✓									
[20]	2022	✓										✓
[21]	2022		✓							✓	✓	
[22]	2022	✓					✓					
[23]	2022	✓		✓								
[24]	2022	✓		✓							✓	
[25]	2022	✓		✓	✓							

alternating optimization algorithm. In [14], the authors have studied the mobile users' offloading process to mobile edge users by using Channel state information and blockchain method. A problem is designed to cope with the problem of decreasing power and time overhead as a Markov Decision problem which is later resolved with a decision tree algorithm and double Q learning methods. The authors in [15] have efficiently managed the resources through federated learning where when user devices are not getting signals from small base stations are connected to macro base stations in digital twin technology. A deep reinforcement learning base scheme is used to cope with the problem of base station assignment and resource allocation. Table 7.1 depicts the literature review comparison of existential research to current search.

7.3 Architecture of UAV in Metaverse

UAVs, as discussed in Sect. 7.2.3, gather attention from multiple perspectives, here, UAV are presented in collaboration with an emerging concept of metaverse. Figure 7.2 shows a general architecture of placement of UAVs, Users, and metaverse server. Providing cellular-based stations is difficult in a situation where catastrophic conditions occur, instead of relying on infrastructure-based technology, we have proposed an infrastructure-less technology where UAV provides a wireless connection to users. As metaverse needs all-time internet connectivity to perform virtual tasks, UAVs can be the best alternative in such a situation. In the current scenario, a disaster has occurred where users have been affected due to major catastrophic conditions. These scenarios are inspired by the catastrophic events of Turkey and Pakistan in 2022 where the earth quack and floods affected the people and cities that urgent immediate connection to cities by roads and cellular communication became difficult. In the prescribed situation, virtual reality can be used where a digital twin

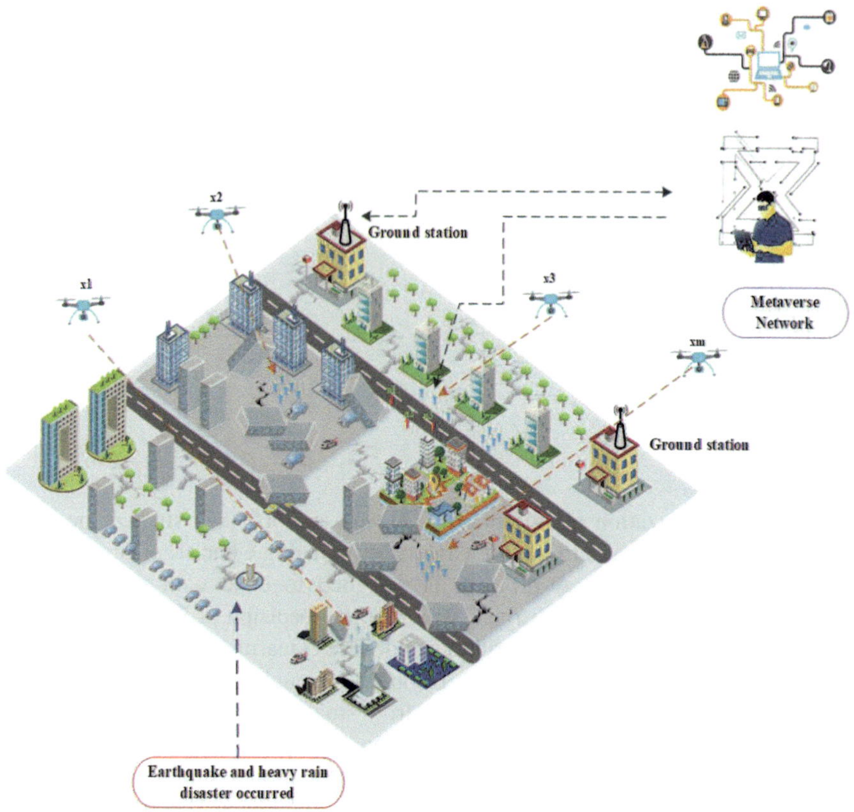

Fig. 7.2 General architecture of metaverse enabled UAV in disaster scenario

of the catastrophic area can also be presented to the rescue agencies to determine what type of help can be needed and will be sent accordingly.

Preliminaries

Before analyzing the system model, some definitions need to be known which will make the system easy to understand. Users are defined by N while the placement of UAVs is shown through M. If m^{th} UAV is placed to provide connection to metaverserver user for medical aid assistance and mirror image, its value will be 1 otherwise 0, and can be written as:

$$x_m = \begin{cases} 1, \text{ if UAV is placed at } m\text{-th location} \\ 0, \text{ otherwise.} \end{cases}$$

In case of metaverse user connection to UAV and UAV connection for medical assistance to ground users, denoted by y_{mn} can be written as:

$$y_{mn} = \begin{cases} 1, \text{ If } m\text{-th UAV is connected to } n\text{-th user} \\ 0, \text{ otherwise.} \end{cases}$$

While β is the percentage of users(on ground and metaverse) who need to be connected, γ_{max} and γ_{min} is the maximum and minimum percentage of users that can be accommodated by UAV.

7.3.1 System Model and Problem Formulation

Figure 7.2 shows a complete UAV user architecture in metaverse environment, Where M number of UAVs and N number of users are placed in disastrous situations. The objective of the provided system is to optimally deploy UAVs and aid users with the assistance of digital twin. Mathematical formulation of the system is as follows;

$$\min_{\substack{x_m, y_{mn} \\ \forall n,m}} \{\frac{\sum_{m=1}^{M} x_m}{M}; \frac{-\sum_{n=1}^{N}\sum_{m=1}^{M} y_{mn}}{N}\} \tag{7.1}$$

Subject to;

$$\begin{aligned} &C1: \sum_{n=1}^{N}\sum_{m=1}^{M} y_{mn} \geq \beta N \\ &C2: y_{mn} \leq x_m, \ \forall n, m \\ &C3: \sum_{m=1}^{M} x_m \geq 1 \\ &C4: \sum_{n=1}^{N} y_{mn} \leq \gamma^{max} x_m \ \forall m \\ &C5: \sum_{n=1}^{N} y_{mn} \geq \gamma^{min} x_m \ \forall m \\ &C6: \sum_{m=1}^{M} y_{mn} \leq 1, \ \forall n \\ &C7: x_m \in \{0, 1\}, \ y_{mn} \in \{0, 1\} \ \forall m, n \end{aligned} \tag{7.2}$$

The min-max problem in 7.1 is an integer linear program that minimizes the number of UAVs and maximizes the number of connections. The proposed model follows some system constraints. C1 is a user connection constraint that supports the users' connectivity needs to be maintained in the system. This constraint will help in assuring connections to all UAV, i.e., proper utilization of UAV. C2 limits the system with UAV deployment, if a UAV is placed at mth location, then a user can get a connection. If no UAV is placed then no user can get a connection. Following constraint C2, C3 assures that a UAV should be placed all the time to provide connectivity to metaverse servers. C4 and C5 are the UAV capacity constraints that bind the UAV with accompanying maximum and minimum connections with users, i.e., according to the capacity of UAVs, users can get connected to UAV which later helps in proper utilization of UAV, and provides an efficient system. C6 assures that no user will get a connection from more than one UAV. C7 is a binary variable constraint that assures the value of UAV and user-UAV connection can be 1 or 0.

7.4 Proposed Algorithm

In this section, a brief review of Branch and bound algorithm (BB) with a competitive heuristic algorithm is given. First an optimal, i.e., BB is proposed to efficiently deploy UAVs with reliable user connection. Then, a low-complexity heuristic algorithm is proposed.

Optimal Algorithm
BB algorithm is used to solve computationally complex problems having exponential growth in time. These types of problems search for all possible solutions until an optimal solution or required solution is achieved. The strategy of the algorithm is break and conquer, where a problem acts as a tree node and is divided into sub-branches where each branch contains further sub-branches, checking on all the solutions. If the obtained solution is better than the previous solution, the previous solution will be discarded and the obtained solution will act as optimal. The algorithm will continue to run until optimal solution is obtained or fathoming is done.

The algorithm is initialized with network parameters, i.e., N, M, β, γ_{min}, γ_{max}. The basic problem will act as a tree node, while S is the container of all the active nodes. The tree node is divided into several subbranches according to the objective function. Once the search for the solution γ after division into subbranches is started, it will continue until optimal solution is obtained. If a better solution will be obtained, γ will be replaced by γ^*. Algorithm 2 will be used to check if there are any repeating indices, it will fix them and return to Algorithm 1.

Greedy Algorithm
Figure 7.3 shows a low complexity heuristic which is proposed to show the competency of optimal algorithm. The algorithm is initialized with basic network parameters. In the pictorial description, the algorithm will check for constraints C4 and C5. If any UAV is accompanying connections that are more than γ^{max},

Algorithm 1 BB algorithm for UAV deployment, user connectivity

Inputs: N , M , $Beta$, γ_{max} , γ_{min} ,
Initialization:
$O_0 \leftarrow Optimization\ Problem$
$N \leftarrow \{O_0\}$, set of active nodes
$\gamma \leftarrow \infty$ objective function value for incumbent solution
$S^* \leftarrow \Phi$
$S \leftarrow \phi,\ current\ solution$
$k \leftarrow 4,\ number\ of\ subproblems\ for\ each\ node$
while $N \neq \phi$ **do**
  $W \leftarrow select(N)$
  $W \leftarrow N \setminus w$
  $Q \leftarrow split(W, k)$
  for $j \epsilon 1, \ldots, k$ **do**
    $[S, \gamma] \leftarrow solve\_ILP(O_j \epsilon\ O)$
    if $S\ is\ complete$ **then**
      if $\gamma < \gamma^*$ **then**
        $S^* \leftarrow S, \gamma \leftarrow \gamma^*$
      end if
    end if
    if $\gamma < \gamma^*$ **then**
      $[S_2, \gamma_2] \leftarrow RSS(S \epsilon\ S^*)$
      if $S_2\ is\ complete$ **then**
        if $\gamma_2 < (1 - \epsilon)\gamma^*$ **then**
          $S^* \leftarrow S_2, \gamma^* \leftarrow \gamma_2$
        end if
      else
        if $\gamma_2 < \gamma^*$ **then**
          $N \leftarrow S_2$
        end if
      end if
    end if
  end for
end while

Algorithm 2 RSS subroutine

Inputs: S , S^*
Initialization:
$I \leftarrow S^* \oplus S$
$S \leftarrow Fix(S(I))$
$[S_2, \gamma_2] \leftarrow Solve\_ILP(S)$

the connections will be removed due to over-utilization. Similarly, if any UAV is carrying connections that are less than γ^{min}, the connections will be removed due to under-utilization. Finally, weighted utility will be calculated for formulation given in 7.1.

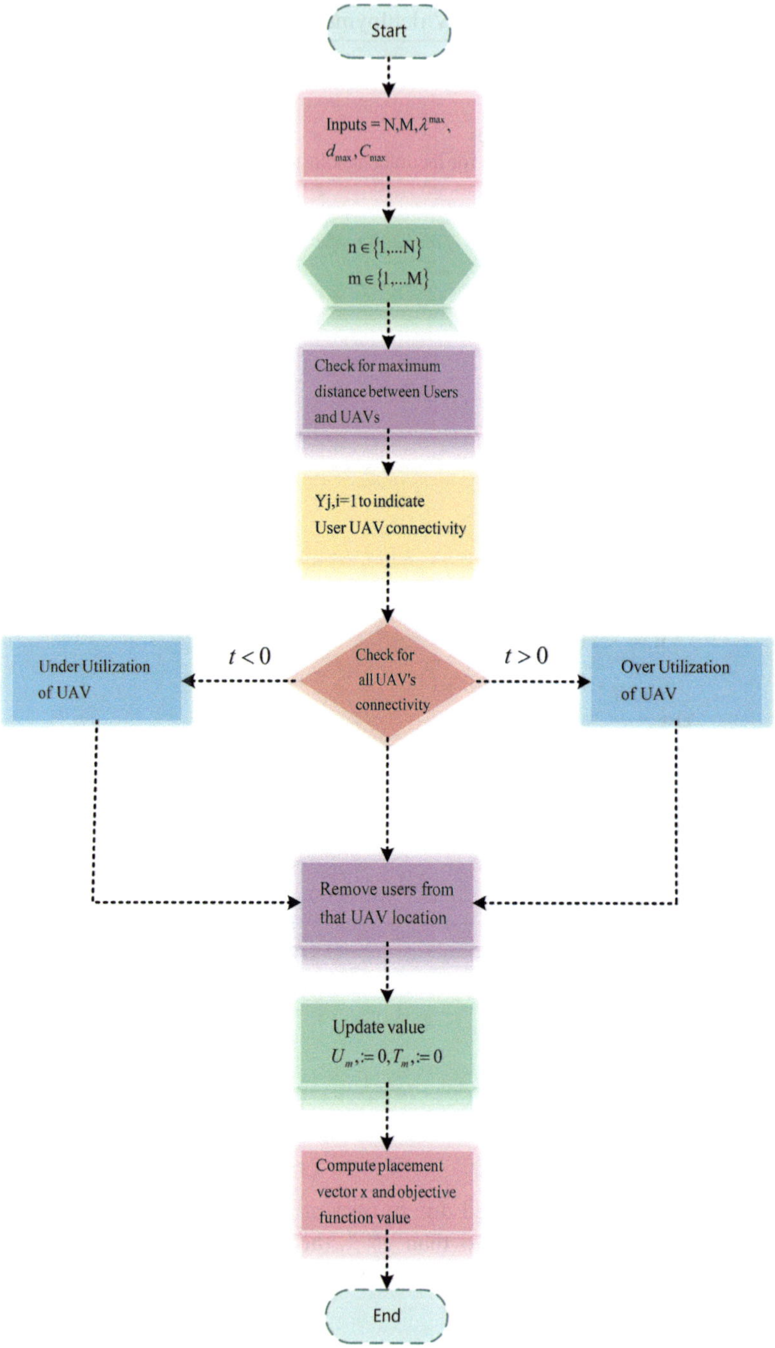

Fig. 7.3 Heuristic algorithm

7.5 Results and Discussions

In this section, the performance of optimal with the greedy algorithm is compared. A scenario is considered where number of users, i.e., N is varied from $N = 50 - 150$, and UAVs are $M = 3 \times 3, 4 \times 4$. The minimum percentage of users need to be connected, i.e., β is varied from $0.4 - 0.7$. Moreover, the defined parameters in C4 and C5, i.e., $\gamma^{max} = 50$, $\gamma_{min} = 10$.

Figure 7.4a, b, is showing the connectivity view of UAV to users at different values of β, i.e., 0.5, and 0.7. As we know from the mathematical formulation in 7.2, β is the minimum percentage of users who need to be connected. However, it can be seen that at 0.5, 81 users out of 150 in the greedy algorithm are connected while at 0.7, 94 users out of 150 are connected which means with an increase in the percentage of connections more users will be accommodated by UAV. The grid is used for the uniform distribution of users. Similarly, in Fig. 7.4, it can be seen that UAV grid size is also increased from 3×3 to 4×4, this increase in grid size will support more users to get connection which is another reason for connectivity of users to UAV. Increasing the grid size allows more users to get connections as 102 connections out of 150 with 4×4 keeping all the parameters same while the 3×3 grid gives 92 out of 150 connections.

Figure 7.5 is an illustration of all the objectives proposed in Eq. 7.1. Since the weighted utility is between installed UAVs and the number of connections. It can be seen that optimal algorithm is competing with greedy algorithm over the range of $\beta = 0.4 - 07$. The competing margin slightly gets lower with an increase in the percentage of users who need to be connected and the size of grid. The overall weighted utility is increased with increased percentage of β and number of installed UAVs, at a smaller value of beta, i.e., 0.4, optimal and greedy algorithms are showing less number of connections to UAV compared to the utility at $\beta = 0.7$. Simulation and results show the performance of the algorithms.

7.6 Summary

Metaverse addresses multiple disciplines which include health sciences, social media applications, official meetings, and many more. In this research, we have given a quick overview of the details of basic metaverse concepts including its elements, components, communication technologies, and enabling technologies. The study has proposed deployment of UAV in disastrous scenarios assisting Digital Twin with urgent connection-making to the relief site. Here, UAVs placement will act as a backbone infrastructure for metaverse communication technology since we are dealing with a disaster scenario, where communication infrastructure got demolished due to catastrophic conditions. In these disastrous scenarios, the digital twin the emerging core technology of metaverse will be helpful by providing a mirror image of the disastrous world to the real world. This image will help the disastrous site in making communication with emergency aid according to the scenario and condition.

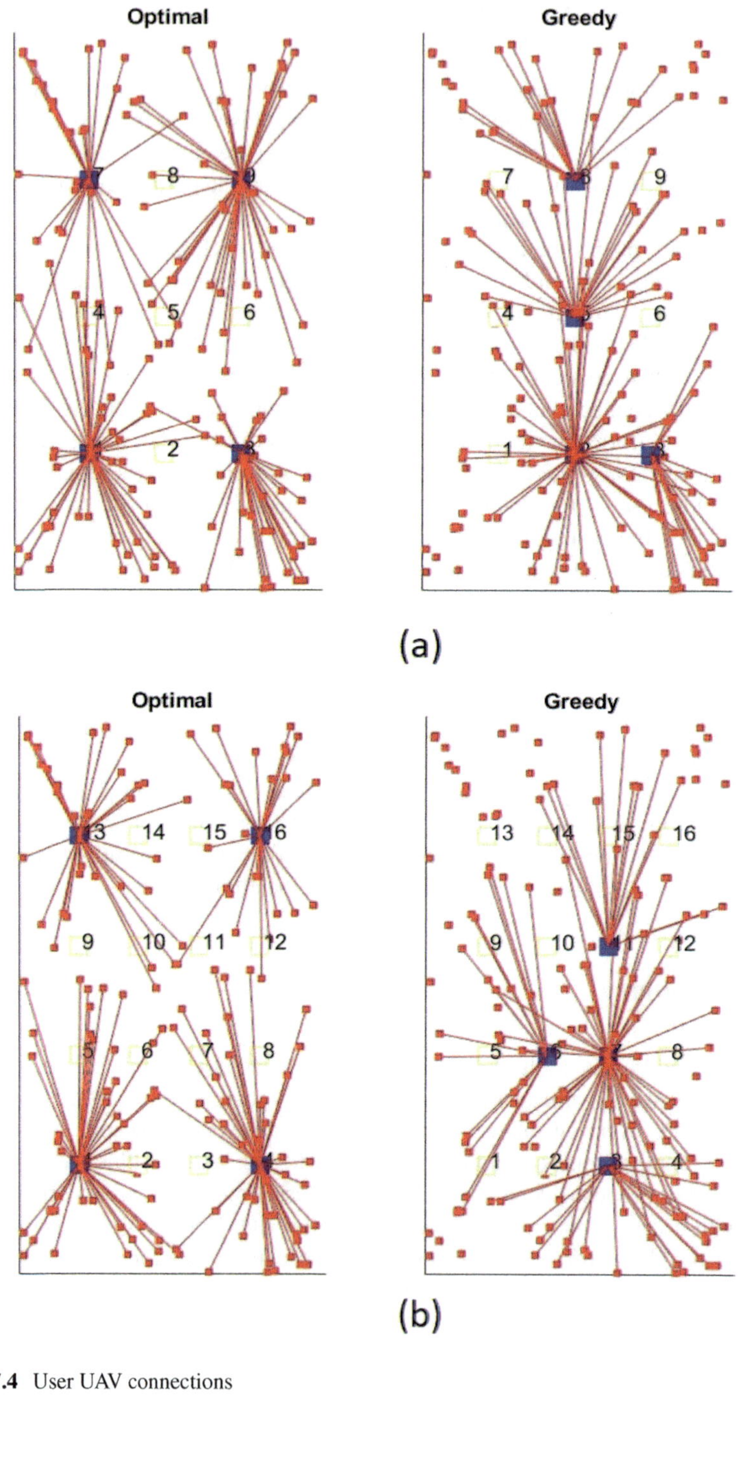

Fig. 7.4 User UAV connections

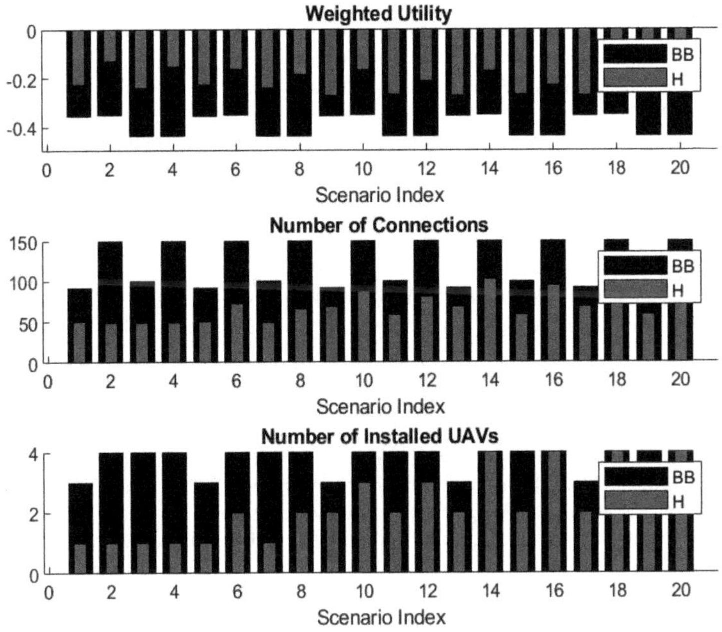

Fig. 7.5 Weighted utility, number of connections, installed UAVs

References

1. S. Chen, D.F Laefer, E. Mangina, State of technology review of civilian UAVs. Recent Patents Eng. **10**(3), 160–174 (2016)
2. S. Lee, Y. Choi, Reviews of unmanned aerial vehicle (drone) technology trends and its applications in the mining industry. Geosyst. Eng. **19**(4), 197–204 (2016)
3. Z. Xiaoning, Analysis of military application of UAV swarm technology, in *2020 3rd International Conference on Unmanned Systems (ICUS)* (IEEE, Piscataway, 2020), pp. 1200–1204
4. S.H. Alsamhi, F. Afghah, R. Sahal, A. Hawbani, M.A. Al-qaness, B. Lee, M. Guizani, Green internet of things using UAVs in b5g networks: a review of applications and strategies. Ad Hoc Netw. **117**, 102505 (2021)
5. R. Shahzadi, M. Ali, H.Z. Khan, M. Naeem, UAV assisted 5g and beyond wireless networks: a survey. J. Netw. Comput. Appl. **189**, 103114 (2021)
6. I.A. Elnabty, Y. Fahmy, M. Kafafy, A survey on UAV placement optimization for UAV-assisted communication in 5g and beyond networks. Phys. Commun. **51**, 101564 (2022)
7. P. Almasan, M. Ferriol-Galmés, J. Paillisse, J. Suárez-Varela, D. Perino, D. López, A.A.P. Perales, P. Harvey, L. Ciavaglia, L. Wong, et al., Network digital twin: context, enabling technologies and opportunities (2022). arXiv:2205.14206
8. O. Hashash, C. Chaccour, W. Saad, Edge continual learning for dynamic digital twins over wireless networks (2022). arXiv:2204.04795
9. D.-Y. Jeong, M.-S. Baek, T. B. Lim, Y.-W. Kim, S.-H. Kim, Y.-T. Lee, W.-S. Jung, I.-B. Lee, Digital twin: technology evolution stages and implementation layers with technology elements. IEEE Access. **10**, 52609–52620 (2022)

10. O. Chukhno, N. Chukhno, G. Araniti, C. Campolo, A. Iera, A. Molinaro, Placement of social digital twins at the edge for beyond 5g iot networks. IEEE Internet Things J. **12**, 1–1 (2022)
11. X. Xu, Z. Liu, M. Bilal, S. Vimal, H. Song, Computation offloading and service caching for intelligent transportation systems with digital twin. IEEE Trans. Intell. Transp. Syst. **23**(11), 20757–20772 (2022)
12. T. Do-Duy, D. Van Huynh, O.A. Dobre, B. Canberk, T.Q. Duong, Digital twin-aided intelligent offloading with edge selection in mobile edge computing. IEEE Wireless Commun. Lett. **11**(4), 806–810 (2022)
13. D. Van Huynh, V.-D. Nguyen, S.R. Khosravirad, V. Sharma, O.A. Dobre, H. Shin, T.Q. Duong, Urllc edge networks with joint optimal user association, task offloading and resource allocation: a digital twin approach. IEEE Trans. Commun. **70**(11), 7669–7682 (2022)
14. T. Liu, L. Tang, W. Wang, Q. Chen, X. Zeng, Digital-twin-assisted task offloading based on edge collaboration in the digital twin edge network. IEEE Internet Things J. **9**(2), 1427–1444 (2021)
15. Y. He, M. Yang, Z. He, M. Guizani, Resource allocation based on digital twin-enabled federated learning framework in heterogeneous cellular network. IEEE Trans. Vehicular Technol. **72**(1), 1149–1158 (2022)
16. Y. Wang, J. Zhao, Mobile edge computing, metaverse, 6g wireless communications, artificial intelligence, and blockchain: survey and their convergence (2022). arXiv:2209.14147
17. F.Y. Wang, R. Qin, X. Wang, B. Hu, Metasocieties in metaverse: metaeconomics and metamanagement for metaenterprises and metacities. IEEE Trans. Comput. Soc. Syst. **9**(1), 2–7 (2022)
18. S.M. Park, Y.G. Kim, A metaverse: taxonomy, components, applications, and open challenges. IEEE Access **10**, 4209–4251 (2022)
19. Y. Wang, Z. Su, N. Zhang, R. Xing, D. Liu, T.H. Luan, X. Shen, A survey on metaverse: fundamentals, security, and privacy. IEEE Commun. Surveys Tutorials. **25**(1), 319–352 (2022)
20. P. Si, J. Zhao, H. Han, K.Y. Lam, Y. Liu, Resource allocation and resolution control in the metaverse with mobile augmented reality (2022). arXiv:2209.13871
21. L.U. Khan, Z. Han, D. Niyato, E. Hossain, C.S. Hong, Metaverse for wireless systems: vision, enablers, architecture, and future directions (2022). arXiv:2207.00413
22. X. Huang, W. Zhong, J. Nie, Q. Hu, Z. Xiong, J. Kang, T.Q. Quek, Joint user association and resource pricing for metaverse: distributed and centralized approaches (2022). arXiv:2208.06770
23. Z. Huang, V. Friderikos, Mobility aware optimization in the metaverse (2022). arXiv:2209.08899
24. T.J. Chua, W. Yu, J. Zhao, Resource allocation for mobile metaverse with the internet of vehicles over 6g wireless communications: a deep reinforcement learning approach (2022). arXiv:2209.13425
25. T. Maksymyuk, J. Gazda, G. Bugár, V. Gazda, M. Liyanage, M. Dohler, Blockchain-empowered service management for the decentralized metaverse of things. IEEE Access **10**, 99025–99037 (2022)

Chapter 8
Digital Twin for UAVs: Architecture, Framework, Challenges and Solutions

Muhammad Amjad

8.1 Introduction

Unmanned Aerial Vehicles (UAVs), have gained much attention in academia and industry due to their applications in civil, military, agriculture, and surveillance [1, 2]. UAVs in conjunction with other state-of-the-art technologies such as artificial intelligence (AI), Internet of Everything (IoE), and Blockchain have achieved the challenging operations such as remote monitoring, teleoperation, aerial photography, and precision landing. These challenging tasks require a framework that could be used to troubleshoot, manage, and optimize the futuristic UAVs. For this purpose, the concept of digital twins for the UAVs has been proposed to optimize and manage the challenging operations of the UAVs for their diverse applications [3]. Digital twins are the virtualization of the physical UAVs operation and processes that can interact with them while monitoring the live operations. Digital twins are the combination of the virtual space, physical space, and the interaction between them.

To perform the more advanced and challenging tasks, the intelligent algorithms are employed in UAVs. The optimization of these intelligent algorithms requires the repeated verification and training of them. To achieve the complex training and testing of these intelligent algorithms for the UAVs, a diverse range of simulation platforms have been proposed [4]. As the physical testing and training incurs cost, effort, and risk, therefore simulation platforms have been adopted. However, the training and testing of the intelligent algorithms for the UAVs using different simulation platforms is very limited and based on the preset factors and environment. To address this limitation, the concept of digital twin for UAVs has been used for the training and testing of the UAVs under different sometimes

M. Amjad (✉)
School of Engineering, Computing and Mathematical Sciences, University of Wolverhampton, Wolverhampton, UK
e-mail: m.amjad6@wlv.ac.uk

M. K. Afzal et al. (eds.), *Digital Twins for Wireless Networks*,
https://doi.org/10.1007/978-3-031-73679-7_8

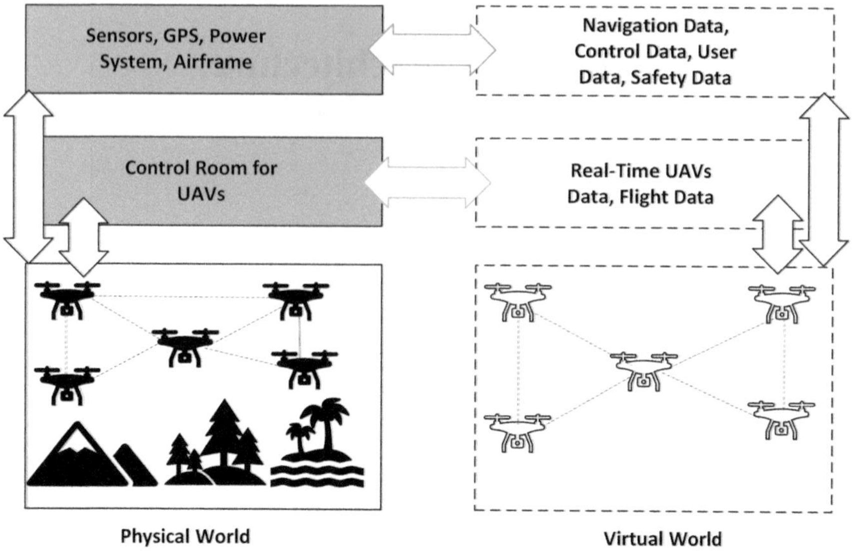

Fig. 8.1 Architecture for digital twins for UAVs operation consisting of the physical world and the virtual world

unknown environments. The digital twin technology could use the live data from the live operation of the UAVs and then adjust the parameters of its processes [5].

As mentioned above, the digital twin for the UAVs consists of the physical layer (world), virtual layer (world), and the interaction between them. The core architecture for the digital twin for the UAVs has been shown in Fig. 8.1. The physical world of the digital twin consists of the physical UAVs (commonly known as drones), wireless communication links among them, control room or platform, and the various drone components. The UAVs or drones are of different kinds depending on their platforms, wings, landing, size, and their supported applications. The wireless links among them are commonly use the advanced wireless communication paradigms such as 5G and beyond 5G. On the other hand, the virtual layer consists of the virtual representation of the drones, and the real-time data related to all or desired processes, such as navigation, flight operations, control, and safety. The operations of the UAVs can be optimized using the real-time data from the UAVs or drones in the virtual world, without interrupting the UAVs physical operations. This advantage makes the UAVs to achieve challenging tasks while operating within limited resources.

In this chapter, a detailed discussion of digital twin-enabled UAVs has been provided while focusing on the enabling architectures and frameworks. To emphasize the importance of UAVs, the diverse application of UAVs in agriculture, security and surveillance, civil, and military have been examined in detail. Different digital twin frameworks for various case studies of UAVs such as teleoperation, edge computing, and vehicular networks have also been presented. Open issues, challenges, and future research directions related to the emerging field of the digital twin enabled UAVs have also been highlighted. In the last, this chapter has been concluded with the main findings and some higher level insights.

8.2 Applications of UAVs

UAVs or drones can be controlled remotely or can fly independently using the intelligent algorithms with proper testing and training. Their applications are of wide-ranging, from civil and agriculture to the military. Initially, the UAVs were only used for defense and military purposes. Now, they are being employed in a diverse range of domains from delivery to pesticide spraying. With the advent of remote sensing and monitoring with UAVs, this technology has been widely accepted as a breakthrough for different complex operations. Below is the detailed discussion of the various applications of UAVs in supporting the different operations/systems.

8.2.1 Agricultural Applications

Agriculture is the core sector aiming to increase growth and productivity by 70% by 2050. This ambitious target can only be achieved with the incorporation of advanced technology and novel methods in the agriculture world. Remote sensing and monitoring has enabled UAVs to be used in this domain. UAVs may be deployed in the agricultural field for various purposes such as weed control, livestock monitoring, pesticide spraying, and irrigation purposes. [6]. Through the applications of UAVs in agriculture, "precision agriculture" can increase yield across many fields and satisfy the future demands of the growing population [7].

8.2.2 Security and Surveillance

With the advances in machine learning and image processing schemes, UAVs have been extensively used for security and surveillance purposes. UAVs, with the help of computer vision and machine learning, have been investigated for district surveillance [8]. For this purpose, the high-resolution images and videos are acquired using on-board sensor technology. In this UAV-aided computer vision scheme, significant progress has been achieved in the domain of structural dynamics. Next generation security and surveillance with help of UAVs are discussed in [9]. In this study, the algorithms are developed for the UAVs autonomous flight operation and localization. To enhance the security operation, an accurate sensor network is also employed.

8.2.3 Civil Applications

In the beginning, the UAVs were only used for defense purposes. However, drones have been extensively used for various civil applications such as transport monitoring, gas and oil pipelines inspections, delivery of goods, and in construction [10, 11]. When UAVs are employed in the civil space, they provide different features such as seamless wireless connectivity to remote users, remote sensing, and inspection of distant gas and pipeline infrastructure.

8.2.4 Military Applications

Contemporary wars are "drones wars". Different types of drones with different capabilities are used in the military field. For example, a quadrotor UAV [12] with an enabled digital twin model has been specifically developed for the military operations. In this work, the mission requirements for single quadrotor UAVs are taken into consideration and the digital twin model is created based on the virtual military field. Digital twins for UAVs have now been extensively studied for their applications in the military field [13]. As mentioned above, initially, the UAVs were developed for military purposes, however, afterwards civil and other applications have also been supported with this technology.

8.3 Different Digital Twins Framework for UAVs Case Studies

Digital twins are the virtual representations of the physical world that can be used for online learning and optimization without interrupting the physical processes. Digital twins for the UAVs, also enable online learning and optimization of the intelligent algorithms and other operations in the virtual domain. Through this technique, seamless and smooth operation of the UAVs is made possible without interrupting the critical operations of the UAVs. Below are some of the use cases, where digital twins for UAVs aid major UAV operations.

8.3.1 Digital Twin-Enabled UAV Teleoperation

Where it is not safe or feasible for humans to operate or control those systems, teleoperation is used to remotely control the systems. UAVs have been extensively used in different teleoperation use cases. The digital twin framework for UAVs to accomplish the complex teleoperation has been proposed in various studies. For

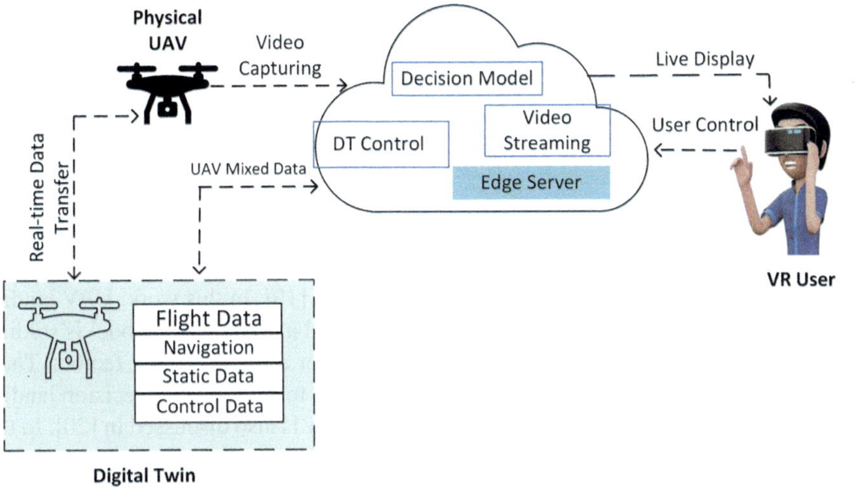

Fig. 8.2 Architecture for digital twins for UAVs operation for teleoperation use case

example, in [14], reliable immersive teleoperation has been proposed through digital twin-enabled UAVs. Figure 8.2, shows the digital twin-enabled UAV framework for the teleoperation procedure [14]. The UAVs are controlled through the cellular networks. A replica of the physical world is created using the digital twin and is equipped with features that are absent in the physical world to investigate the critical issues such as latency and reliability. AeroVR is the digital twin enabled UAVs framework for achieving the teleoperation with tactile feedback [15]. This use case is specifically designed for the operations such as emergency and e-commerce.

8.3.2 UAV-Enabled Mobile Edge Computing System

Through the mobile edge computing (MEC), the computational tasks can be brought very close to the mobile users. A digital twin-enabled UAV framework has been developed to achieve MEC or multi-access edge computing. Resource allocation for the efficient operation of a large number of IoT devices requires a UAV fleet with a digital twin framework to be deployed, while considering the computing and trajectory constraints [16]. A tasks offloading scheme has been investigated within UAV-enabled MEC while considering the digital twin platform in [17]. In this work, the energy consumption by the MEC system has been optimized while considering the constrains of the UAVs trajectories, power, and processing delay. Another work that considers the UAV-enabled MEC use case with a digital twin framework, focusing on latency and reliability for the 6G network, is presented in [18]. A beyond-5G use case of ultra reliable and low latency communications (uRLLC), with stringent constrains on delay and reliability, has been investigated in UAVs enabled MEC with the digital twin framework.

8.3.3 UAV Precision Landing

Landing of UAVs at the accurate point with the correct position is critical for the successful operation of the UAVs. Precision landing of UAVs using its digital twin model to optimize landing has been investigated in [19]. In this work, UAV landing with confidence in the pre-defined environment and a digital twin model is studied. Landing errors occur due to sensing errors, operator error, and other factors. These factors are optimized in the digital twin framework to achieve the precision landing for the UAVs. UAV landing with enhanced accuracy is also discussed in [20]. In this work, a two-level framework consisting of data-driven and digital twin approaches proposes the digital twin instance for landing of the UAVs.

8.3.4 UAV Security Enhancement Using Digital Twin

Securing UAVs and the enabling digital twin framework from various security attacks is recognized as the core aspect for developing the UAV-enabled digital twin framework. Various studies have been conducted to provide a secure framework for UAVs. In [21], a secure computing approach is introduced to protect UAVs from the different security threats. In this approach, two secure methods, namely secure communications and secure resource allocations for UAVs have been investigated. A digital twin framework to secure UAV operations is also investigated in [22]. Different machine learning approaches are utilized in the digital twin framework to detect cyber intrusions and anomalies.

8.3.5 UAV Assisted Vehicular Technology

The digital twin-enabled UAV framework for the vehicular communications has provided the opportunity to achieve the seamless connectivity, higher throughput, and massive connectivity for the vehicular technology networks (VANETs). Reconfigurable intelligent surfaces are the new domains for handling the randomness created from the propagation environment. UAVs and RIS are promising technologies for VANETs with their potential applications in smart cities [23]. UAVs-RIS with the digital twin model has been studied to understand and optimize the randomness, prediction, and mobility models of the VANETs. Figure 8.3 shows the UAVs-RIS enabled digital twin framework for the vehicular networks. An approach named FlexEdge [24], has been proposed to introduce the digital twin-enabled UAV

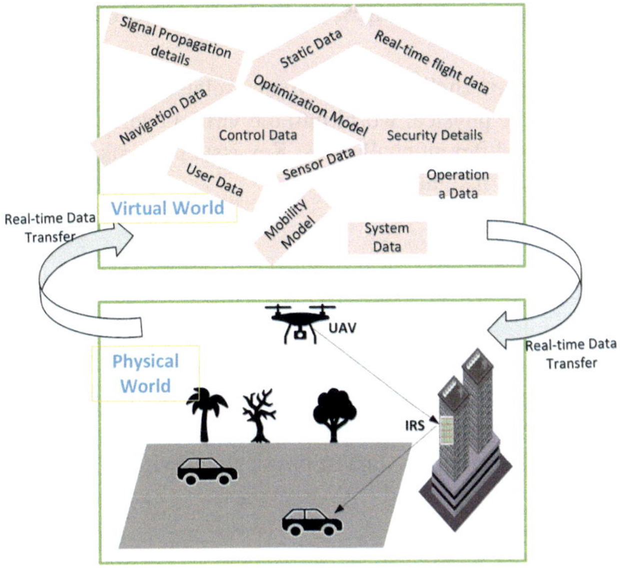

Fig. 8.3 Architecture for digital twins for UAVs-RIS for the vehicular communications

framework for the vehicular edge computing. Through the digital twin framework, energy minimization approach for vehicular edge computing has been proposed and then energy usage is optimized. A resource allocation scheme for digital twin-enabled UAVs in the Internet of vehicles network has also been studied in detail in [25]. Resource allocation, more specifically the computational resources, has been explored in the virtual layer of the digital twins to achieve the seamless connectivity in the Internet of vehicles network.

8.3.6 *Predictive Maintenance of UAVs Using Digital Twins*

Predictive maintenance using the UAV-enabled digital twin framework has gained much attention due to harsh conditions where humans cannot reach. In scenarios, where it is not feasible for humans to perform maintenance tasks, UAVs are used with some controlled or autonomous flight. Digital twin frameworks for UAVs used for maintenance purposes have been investigated while considering the virtual and physical layers. In [26], real-time fault detection and identification using the UAV-enabled digital twin framework are investigated in detail. This has been achieved using optical strain sensors and fault detection algorithms. A machine learning approach via predictive modeling to design the optimal sensor placement in the digital twin framework is studied in [27]. This sensor placement strategy can then be utilized for the UAVs.

8.4 Open Issues, Challenges, and Future Research Direction

There are a plethora of challenges and issues related to the digital twin-enabled UAVs. Replicating the physical world into the virtual one, with complex details of the UAVs introduces different challenges. Key issues are discussed in detail below with some higher-level insights.

8.4.1 Complex Data Handling

UAVs produce a lot of real-time data from various flight trajectories, sensors, actuators, motors, and navigation systems. This data must be timely and securely transmitted to the virtual domain and synchronized. Handling and performing computation using this complex data in physical and virtual layers poses a serious challenge for digital twin-enabled UAV networks.

8.4.2 Avoiding Security Risks

Securing the digital twin framework for UAVs is a critical challenge considering the importance of UAVs different challenging applications. For this purpose, a well-established security paradigm is needed to counter intruders and different security threats for the smooth operation of digital twin-enabled UAVs.

8.4.3 Operational Cost

The digital twin footprint requires the sophisticated training and awareness to handle the overall structure. The operational cost involved in running this framework is a serious challenge. Considering the higher prices of drones, the operational cost of their digital twin component will incur a serious burden in exploiting the potential of digital twin-enabled UAVs.

8.4.4 AI and Machine Learning

Intelligent algorithms have been utilized in UAVs to achieve the various challenging tasks. For this purpose, AI and machine learning algorithms need the proper training and testing in real-time. However, considering the resource constraints and complexity of the various machine learning algorithms, the training and testing in a virtual environment sometimes pose a challenge.

8.4.5 Compatibility Issues

Various components of the UAVs usually come from different vendors and follow different standards paradigm. This introduces a compatibility issue between the physical world and the software used in the virtual world. Compatibility is a serious concern for the smooth sharing of data between the virtual world and the physical world.

The current version of the digital twin framework for the UAVs has a limited capabilities in terms of computation, security, and communications. As future work, enhanced capabilities, more security, and reliable communications paradigms are required for the smooth operation of digital twin-enabled UAVs.

8.5 Conclusion

UAVs are a promising technology that has been deployed in various fields such as military, agriculture, civil, and the medical industry. To achieve the smooth operations of the UAVs, online learning and self-optimization in UAVs can be achieved with the help of digital twin technology. In this chapter, a critical review of the digital twin-enabled UAVs framework is provided. Different state-of-the-art applications of the UAVs were discussed in more detail. Different use cases of the digital twin enabled UAVs such as teleoperation, vehicular networks, maintenance, and security with their respective architecture were also investigated. Some key challenges and issues with high-level insights were highlighted with key future research directions.

References

1. N. Sarantinoudis, N. Vitzilaios, G. Arampatzis, Applications of digital twins in UAVs, in *2024 International Conference on Unmanned Aircraft Systems (ICUAS)* (2024), pp. 450–457
2. M. Singh, R. Srivastava, E. Fuenmayor, V. Kuts, Y. Qiao, N. Murray, D. Devine, Applications of digital twin across industries: a review. Appl. Sci. **12**(11), 5727 (2022)
3. Z. Lv, D. Chen, H. Feng, R. Lou, H. Wang, Beyond 5G for digital twins of UAVs. Comput. Netw. **197**, 108366 (2021)
4. Y. Yang, W. Meng, H. Li, R. Lu, M. Fu, A digital twin platform for multi-rotor UAV, in *2021 40th Chinese Control Conference (CCC)* (IEEE, Piscataway, 2021), pp. 7909–7913
5. Y. Yang, W. Meng, S. Zhu, A digital twin simulation platform for multi-rotor UAV, in *2020 7th International Conference on Information, Cybernetics, and Computational Social Systems (ICCSS)* (IEEE, Piscataway, 2020), pp. 591–596
6. D. Rakesh, N.A. Kumar, M. Sivaguru, K.V.R. Keerthivaasan, B.R. Janaki, R. Raffik, Role of UAVs in innovating agriculture with future applications: a review, in *2021 International Conference on Advancements in Electrical, Electronics, Communication, Computing and Automation (ICAECA)* (IEEE, Piscataway, 2021), pp. 1–6

7. D.C. Tsouros, S. Bibi, P.G. Sarigiannidis, A review on UAV-based applications for precision agriculture. Information **10**(11), 349 (2019)
8. S. Saha, S. Banerjee, S. Dhar, A. Bandyopadhyay, S. Saha, S. Ray, Unmanned aerial vehicle for district surveillance with computer vision and machine learning, in *2023 7th International Conference on Electronics, Materials Engineering & Nano-Technology (IEMENTech)* (IEEE, Piscataway, 2023), pp. 1–6
9. M.F. Ayub, F. Ghawash, M.A. Shabbir, M. Kamran, F.A. Butt, Next generation security and surveillance system using autonomous vehicles, in *2018 Ubiquitous Positioning, Indoor Navigation and Location-Based Services (UPINLBS)* (IEEE, Piscataway, 2018), pp. 1–5
10. H. Shakhatreh, A.H. Sawalmeh, A. Al-Fuqaha, Z. Dou, E. Almaita, I. Khalil, N.S. Othman, A. Khreishah, M. Guizani, Unmanned aerial vehicles (UAVs): a survey on civil applications and key research challenges. IEEE Access **7**, 48572–48634 (2019)
11. A. Otto, N. Agatz, J. Campbell, B. Golden, E. Pesch, Optimization approaches for civil applications of unmanned aerial vehicles (UAVs) or aerial drones: a survey. Networks **72**(4), 411–458 (2018)
12. G. Ji, J.G. Hao, J.L. Gao, C.Z. Lu, Digital twin modeling method for individual combat quadrotor UAV, in *2021 IEEE 1st International Conference on Digital Twins and Parallel Intelligence (DTPI)* (IEEE, Piscataway, 2021), pp. 1–4
13. A.F. Mendi, T. Erol, D. Doğan, Digital twin in the military field. IEEE Internet Comput. **26**(5), 33–40 (2021)
14. N. Sehad, X. Tu, A. Rajasekaran, H. Hellaoui, R. Jäntti, M. Debbah, Towards enabling reliable immersive teleoperation through digital twin: a UAV command and control use case, in *GLOBECOM 2023–2023 IEEE Global Communications Conference* (IEEE, Piscataway, 2023), pp. 6420–6425
15. G.A. Yashin, D. Trinitatova, R.T. Agishev, R. Ibrahimov, D. Tsetserukou, Aerovr: virtual reality-based teleoperation with tactile feedback for aerial manipulation, in *2019 19th International Conference on Advanced Robotics (ICAR)* (IEEE, Piscataway, 2019), pp. 767–772
16. S. Shen, Y. Fu, K. Yang, Joint resource allocation in digital-twin-assisted and UAV-enabled mobile edge computing system, in *2023 IEEE 3rd International Conference on Electronic Technology, Communication and Information (ICETCI)* (IEEE, Piscataway, 2023), pp. 302–306
17. B, Li, Y. Liu, L. Tan, H. Pan, Y. Zhang, Digital twin assisted task offloading for aerial edge computing and networks. IEEE Trans. Veh. Technol. **71**(10), 10863–10877 (2022)
18. T.Q. Duong, D.V. Huynh, Y. Li, E. Garcia-Palacios, K. Sun, Digital twin-enabled 6g aerial edge computing with ultra-reliable and low-latency communications, in *2022 1st International Conference on 6G Networking (6GNet)* (IEEE, Piscataway, 2022), pp. 1–5
19. J. Uddin, M.F. Wadud, R. Ashrafi, M.G.R. Alam, M.K. Rhaman, Landing with confidence: the role of digital twin in UAV precision landing, in *2023 10th International Conference on Recent Advances in Air and Space Technologies (RAST)* (IEEE, Piscataway, 2023), pp. 1–6
20. A. McClellan, J. Lorenzetti, M. Pavone, C. Farhat, A physics-based digital twin for model predictive control of autonomous unmanned aerial vehicle landing. Philos. Trans. R. Soc. A **380**(2229), 20210204 (2022)
21. S. Park, M. Yoo, S. Jung, M. Choi, J. Kim, Introduction to secure computing methods in unmanned aerial vehicles, in *2022 Thirteenth International Conference on Ubiquitous and Future Networks (ICUFN)* (IEEE, Piscataway, 2022), pp. 239–241
22. B. Fraser, S. Al-Rubaye, S. Aslam, A. Tsourdos, Enhancing the security of unmanned aerial systems using digital-twin technology and intrusion detection, in *2021 IEEE/AIAA 40th Digital Avionics Systems Conference (DASC)* (IEEE, Piscataway, 2021), pp. 1–10
23. M. Wu, Y. Xiao, Y. Gao, M. Xiao, Digital Twin for UAV-RIS Assisted Vehicular Communication Systems. IEEE Trans. Wireless Commun. **23**(7), 7638–7651 (2024)
24. B. Li, W. Xie, Y. Ye, L. Liu, Z. Fei, Flexedge: digital twin-enabled task offloading for UAV-aided vehicular edge computing. IEEE Trans. Veh. Technol. **72**(8), 11086–11091 (2023)
25. B. Hazarika, K. Singh, C.P. Li, A. Schmeink, K.F. Tsang, RADiT: resource allocation in digital twin-driven UAV-aided internet of vehicle networks. IEEE J. Sel. Areas Commun. **41**, 12 (2023)

26. P.C. Berri, C. Corsi, M.D. Dalla Vedova, A.D. Laudani, P. Maggiore, C. Secci, Study of packaging and installation of FBG sensors for monitoring of aircraft systems, in *Proceedings of the 30th European Safety and Reliability Conference and the 15th Probabilistic Safety Assessment and Management Conference* (Research Publishing Services, Singapore, 2020)
27. M.G. Kapteyn, K.E. Willcox, Design of digital twin sensing strategies via predictive modeling and interpretable machine learning. J. Mech. Des. **144**(9), 091710 (2022)

9783031736780